CDMA

Addison-Wesley Wireless Communications Series

The Addison-Wesley Wireless Communications Series provides engineering and computer science practitioners with comprehensive reference books on new and emerging mobile communication technologies. Professionals involved in the design, implementation, maintenance, and management of wireless communications networks will find necessary information on these cutting edge advancements. Dr. Andrew J. Viterbi, a pioneer in this field, will be the Consulting Editor for the series.

CDMA

Principles of
Spread Spectrum
Communication

Andrew J. Viterbi

ADDISON-WESLEY

An imprint of Addison Wesley Longman, Inc.

Reading, Massachusetts   Harlow, England   Menlo Park, California
Berkeley, California   Don Mills, Ontario   Sydney
Bonn   Amsterdam   Tokyo   Mexico City

Many of the designations used by manufacturers and sellers to distinguish their products are claimed as trademarks. Where those designations appear in this book and Addison-Wesley was aware of a trademark claim, the designations have been printed in initial caps or all caps.

The publisher offers discounts on this book when ordered in quantity for special sales. For more information please contact:

Corporate & Professional Publishing Group
Addison-Wesley Publishing Company
One Jacob Way
Reading, Massachusetts 01867

Library of Congress Cataloging-in-Publication Data
Viterbi, Andrew J.
   CDMA : principles of spread spectrum communication / Andrew J. Viterbi.
      p. cm.
   Includes bibliographical references and index.
   ISBN 0-201-63374-4
   1. Code division multiple access. I. Title.
TK5103.45.V57 1995
321.3845—dc20   94-23800
                                                                              CIP

Text design by Wilson Graphics & Design (Kenneth J. Wilson)
Text printed on recycled and acid-free paper

ISBN 0-201-63374-4
4 5 6 7 8 9 10 11 12–MA–00999897
Fourth printing, March 1997

To the Ladies in my Life,
Maria and Lenka, of Blessed Memories,
and
Erna, Audrey, Caryn, Alexandra, and Samantha

CONTENTS

LIST OF FIGURES AND TABLES

Chapter Six

PREFACE

Spread spectrum communication technology has been used in military communications for over half a century, primarily for two purposes: to overcome the effects of strong intentional interference (jamming), and to hide the signal from the eavesdropper (covertness). Both goals can be achieved by spreading the signal's spectrum to make it virtually indistinguishable from background noise. Several texts, or portions of texts, on this subject have been published over the past twenty years. This book is the first to present spread spectrum technology specifically for commercial wireless applications.

In response to an ever-accelerating worldwide demand for mobile and personal portable communications, spread spectrum digital technology has achieved much higher bandwidth efficiency for a given wireless spectrum allocation, and hence serves a far larger population of multiple access users, than analog or other digital technologies. While it is similar in implementation to its military predecessors, the spread spectrum wireless network achieves efficiency improvements by incorporating a number of unique features made possible by the benign noise-like characteristics of the signal waveform. Chief among these is universal frequency reuse (the fact that all users, whether communicating within a neighborhood, a metropolitan area, or even a nation, occupy a common frequency spectrum allocation). Besides increasing the efficiency of spectrum usage, this also eliminates the chore of planning for different frequency allocation for neighboring users or cells. Many other important multiple access system features are made possible through this universal frequency reuse by terminals employing wideband (spread) noise-like signal waveforms. Most important is fast and accurate power control, which ensures a high level of transmission quality while overcoming the "near-far" problem by maintaining a low transmitted power level for each terminal, and hence a low level of interference to other user terminals. Another is mitigation of faded transmission through the use of a Rake receiver, which constructively combines multipath components rather than allowing them to destructively combine as in narrowband transmission. A third major benefit is soft handoff among multiple cell base stations, which provides improved cell-boundary performance and prevents dropped calls.

In Chapters 2 to 5, this book covers all aspects of spread spectrum transmission over a physical multiple-access channel: signal generation, synchronization, modulation, and error-correcting coding of direct-sequence spread spectrum signals. Chapter 6 relates these physical layer functions to link and network layer properties involving cellular coverage, Erlang capacity, and network control. This outline is unusual in bringing together several wide-ranging technical disciplines, rarely covered in this sequence and in one text. However, the presentation is well integrated by a number of unifying threads. First, the entire text is devoted to the concept of universal frequency reuse by multiple users of multiple cells. Also, two fundamental techniques are used in a variety of different forms throughout the text. The first is the finite-state machine representation of both deterministic and random sequences; the second is the use of simple, elegant upper bounds on the probabilities of a wide range of events related to system performance.

However, given the focus on simultaneous wideband transmission for all users over a common frequency spectrum, the text purposely omits two important application areas: narrowband modulation and coding methods, including multipoint signal constellations and trellis codes; and frequency hopped multiple access, where modulation waveforms are instantaneously narrowband over the duration of each hop. It also generally avoids digressions into principles of information theory. In short, although the material covered through Chapter 5 mostly also applies to narrowband digital transmission systems, the book mainly covers topics that apply to wideband spread spectrum multiple access.

Three motivating forces drove me to write this book. It began with my three decades of teaching within the University of California system. There, keeping with the healthy trend in communication engineering courses, I tried to make theory continually more pertinent to applications. Then there was the fulfillment of a voluntary commission for the Marconi Foundation, which honored me with a Marconi Fellowship award in 1990. Most important was my participation in a significant technological achievement in communication system evolution: the implementation, demonstration, and standardization of a digital cellular spread spectrum code-division multiple access (CDMA) system. Adopted in 1993 by the Telecommunication Industry Association, the CDMA standard IS-95 is the embodiment of many of the principles presented in this text. Although this book is not meant solely for this purpose, it does explain and justify many of the techniques contained in the standard. I emphasize, however, that my goal is to present the principles underlying spread spectrum communication, most but not all of which apply to this standard. It is not to describe in detail how the principles were applied. This is left to the practicing engineer with the patience and commitment to delve into the details and correlate them with the principles presented here.

Which brings me to the question of prerequisites for a basic understanding. Several excellent texts on statistical communication and information theory have been available for almost four decades. Thus, I have not tried to provide all the fundamentals. The text is nevertheless self-contained: any significant results are derived either in the text or in appendices to the chapter where they are first used. Still, the reader should have at least an undergraduate electrical engineering background with some probability and communication engineering content. A first-year engineering graduate course in communication theory, stochastic processes, or detection and estimation theory would be preferable. As a text for a graduate-level course, the book can be covered in one semester, and with some compromises even in one quarter. It is equally suitable for a one- or two-week intensive short course.

This leaves only the pleasant task of thanking the many contributors to the creation of this text. First, from my superb group of colleagues at QUALCOMM Incorporated, running the gamut from mature and renowned engineers to newly minted graduates, have come the inventive system concepts and the innovative implementation approaches that turned the complex concepts into a useful reality. Among the major contributors, Klein Gilhousen, Irwin Jacobs, Roberto Padovani, Lindsay Weaver, and Charles Wheatley stand out. On the more focused aspects of the text, and the research which preceded it, I owe an enormous debt to Audrey Viterbi. She contributed not only ideas, but also the considerable dedication to turn fluid concepts and derivations into firmer results with solid theoretical or simulation support. Finally, she was the first to read, critique, and error-correct the entire manuscript. Over a number of years, Ephraim Zehavi's many ideas and novel approaches have produced results included here. Jack Wolf, always a clear expositor, suggested several improvements. When it came to reviewing the final text and offering corrections and changes, I am indebted to more people than I can recall. Foremost among them are my collaborators at QUALCOMM, including Joseph Odenwalder, Yu-Cheun Jou, Paul Bender, Walid Hamdy, Samir Soliman, Matthew Grob, John Miller, and John McDonough. The last three served as experimental subjects among the first set of graduate students on which I class-tested the entire text. Very helpful outside reviews have come from Robert Gallager, Bijan Jabbari, Allen Levesque, James Mazo, Raymond Pickholtz, and Robert Scholtz. To all of the above, and especially to Deborah Casher, my infinitely patient and cooperative assistant, who processed all my words and equations, I express my sincere thanks.

CHAPTER ONE

Introduction

1.1 Definition and Purpose

Spread spectrum modulation, a wireless communication technique, uses a transmission bandwidth many times greater than the information bandwidth or data rate of any user. We denote the bandwidth in hertz by W and the data rate in bits/second by R. The ratio W/R is the bandwidth spreading factor or processing gain. Values of W/R ranging from one hundred to one million (20 dB to 60 dB) are commonplace. Spread spectrum applications fall into several broad categories:

1. high tolerance to intentional interference (jamming) or unintentional interference, through its suppression by an amount proportional to the spreading factor;

2. position location and velocity estimation, with accuracy increasing proportionally to the spreading bandwidth;

3. low detectability of transmitted signal by an unintended receiver, which decreases with increased spreading factor;

4. multiple access communication by a large population of relatively uncoordinated users of a common spectral allocation in the same and neighboring geographical areas, with the number of simultaneous users proportional to the spreading factor.

This text will concentrate on the last application area. It may seem counterintuitive that multiple access through sharing of a common (spread) spectrum can actually compete in spectral efficiency with more traditional multiple access techniques, which isolate users by providing each with unique disjoint frequencies or time slots. Yet this text will demonstrate that this approach can achieve a considerably higher capacity than any other multiple access technique.

1.2 Basic Limitations of the Conventional Approach

Traditionally, each user of a multiple access system is provided with certain resources, such as frequency or time slots, or both, which are disjoint from those of any other user. In this way the multiple access channel reduces to a multiplicity of single point-to-point channels, assuming perfect isolation of each user's transmission resources from those of all other users. Each channel's capacity is limited only by the bandwidth and time allotted to it, the degradation caused by background noise mostly of thermal origin, and propagation anomalies, which produce multipath fading and shadowing effects. Apparently, then, the composite capacity of the individual components of the multiple access channels would be equal to the capacity of a single user occupying the entire composite resources of the individual channels.

This conclusion suffers from three weaknesses. The first is that it assumes that all users transmit continuously for the same length of time. This may be nearly true for packet data transmission, but usually not for circuit-switched voice transmission. In a two-person conversation, each speaker is active less than half the time; when resources are assigned exclusively, reallocation of the channel for brief periods requires rapid circuit switching between the two users. And even if this is possible, short pauses between syllables cannot be rapidly reallocated.

Another important consideration, which more seriously contradicts the conclusion above, is geographical reallocation of spectrum. Until recently, this question applied primarily to broadcasting. A local television broadcaster's VHF or UHF frequency allocation for Los Angeles could not be allocated to either San Diego or Santa Barbara, each less than 200 km away, though it could be reused in San Francisco, more than 500 km away. While broadcasting is a one-to-many channel, with one transmitter serving many users, multiple access is a many-to-one channel, with one terrestrial base station or satellite hub station receiving from multiple users. Until the advent of cellular wireless mobile telephony, multiple access systems usually consisted of one cell. This cell covered either an urban area from a single base station in a high place, or a larger geographical region (sometimes nationwide) from a single satellite transponder accessed by only a few earth terminals. With the arrival of cellular service, frequency reuse became central to realizing multiple access for a much larger user population scattered over many, often contiguous, metropolitan areas. In North America, a single allocation of a large number of channels was made to each of two regional service providers. The same frequency spectrum was allocated for all regions, both urban and rural. The individual regional carriers were left to coordinate usage of the channels within their own regions and with contiguous regions. The analog technology standard, known as Advanced Mobile Phone System (AMPS), employs FM modulation and occupies a 30 kHz frequency channel slot for each voice signal [Lee, 1989]. Allocations of 12.5 MHz each for the base-station-to-mobile transmission (known as forward link or downlink) and for the mobile-to-base-station transmission (known as reverse link or uplink) provided a little more than 400 channels in each direction. Since the demand in urban areas is many times greater than this, many cellular base stations must be employed, located less than 1 km apart in congested areas. The potential mutual interference among neighboring cells requires a more careful allocation of resources than the gross apportionment employed in broadcasting. The idealized allocation of cellular channels is illustrated in Figure 1.1, where the cells are shown as contiguous hexagons. The same frequency or time slots are never reused in contiguous cells, but may be reused by noncontiguous cells. Figure 1.1 shows two contiguous copies of a multiple-cell "cluster," which is reused as a "building block" to form a replicating pattern over an arbitrarily large area. Two cells that employ the same allocation, and hence interfere with each other, are thus separated by more than one cell diameter. This degree of isolation is sufficient because the terrestrial propagation power loss increases according to approximately the fourth power of the distance. Also, sources of interference are reduced by employing sectored antennas at the base station, with each sector using different frequency bands.

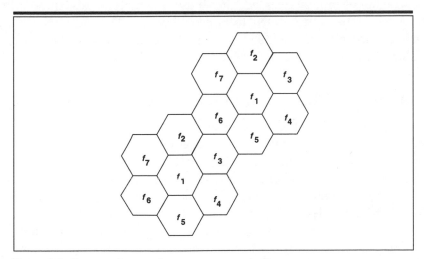

Figure 1.1 Two contiguous frequency reuse clusters.

The consequence, however, is that the number of channels per cell is reduced by the reuse factor (seven in the example of Figure 1.1). Thus, the approximately 400 analog channels for each service provider are reduced to less than 60 per cell. Using sectored antennas, while reducing interference, does not increase the number of slots; for example, each sector of a three-sectored antenna is allocated one-third of the slots (about 20 per sector for the example). Even with these precautions, the interference power received at a given base station from reused channels in other cells is only about 18 dB below the signal power received from the desired user of the same channel in the given cell [Lee, 1989]. Reuse factors (clusters) as low as 4 and even 3 have been considered, but unless more tolerant modulation techniques can be utilized, decreasing the distance between interfering cells increases the other-cell interference to the point of unacceptable signal quality.

A third source of degradation common to all multiple access systems, particularly in terrestrial environments, is multipath. Phase cancellation between different propagation paths can cause severe fading, worsening the effect of interference by lowering the available signal power. This phenomenon is particularly severe when each channel is allocated a narrow bandwidth.

A multiple access system that is more tolerant of interference can be produced by using digital modulation techniques at the transmitter (including both source coding and channel error-correcting coding) and the corresponding signal processing techniques at the receiver. Digital methods by themselves, however, will not change the basic limitations: channels cannot be rapidly released when a user's activity ceases temporarily, frequency reuse is restricted to noncontiguous cells, and narrowband channels are more susceptible to multipath fading. Together these factors reduce the channel capacity by about an order of magnitude.

1.3 Spread Spectrum Principles

A completely different approach, suggested at least in part by the principles of Shannon's information theory, does not attempt to allocate disjoint frequency or time resources to each user. Instead this approach allocates all resources to all simultaneous users, controlling the power transmitted by each to the minimum required to maintain a given signal-to-noise ratio for the required level of performance. Each user employs a noiselike wideband signal occupying the entire frequency allocation for as long as it is needed. In this way, each user contributes to the background noise affecting all the users, but to the least extent possible. This additional

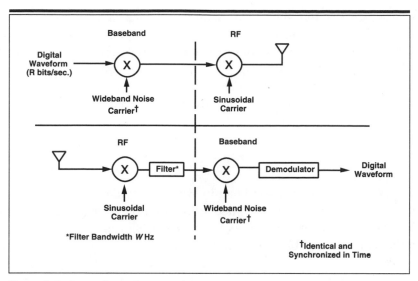

Figure 1.2 Spread spectrum modem.

interference limits capacity, but because time and bandwidth resource allocations are unrestricted, the resulting capacity is significantly greater than for conventional systems.

A coarse estimate of the reverse link capacity[1] achieved by such a spread spectrum system is obtained by the following argument. Suppose each user employs a wideband Gaussian noise carrier. For our purposes, suppose this waveform is stored at both transmitter and receiver, and that the modulation and demodulation are simply multiplication operations at baseband, synchronized between locations (Figure 1.2). Suppose further that each user's transmitted power is controlled so that all signals are received at the base station at equal power levels. If the received signal power of each user is P_S watts, and the background noise is negligible, the total interference power, I, presented to each user's demodulator is

$$I = (k_u - 1)P_S, \tag{1.1}$$

where k_u is the total number of equal energy users.

---

[1] The forward link capacity presents a more complex scenario. Suffice it to say that usually the one-to-many channel is less demanding than the many-to-one reverse channel, even though other-cell interference requires a more complicated analysis, as described in Chapter 6.

Now suppose that the digital demodulator for each user can operate against Gaussian noise at a bit-energy-to-noise-density[2] level of E_b/I_0. This parameter is the figure of merit of the digital modem. It varies typically between 3 and 9 dB, depending on its implementation, use of error-correcting coding, channel impairments such as fading, and, of course, error rate requirements. The noise density received by each user's demodulator is

$$I_0 = I/W, \tag{1.2}$$

since W Hz is the entire spread bandwidth of the wideband noise carriers, whose spectral density is assumed to be uniform over this bandwidth. Similarly, the received energy per bit is the received signal power divided by the data rate R in bits per second, so that

$$E_b = P_S/R. \tag{1.3}$$

Combining (1.1) through (1.3) shows that under these assumptions, the number of simultaneous users able to coexist in an isolated cell is related to the spreading factor and the demodulator's E_b/I_0 requirement by

$$k_u - 1 = \frac{I}{P_S} = \frac{W/R}{E_b/I_0}. \tag{1.4}$$

Suppose further that two more processing features are added to the spread spectrum multiple access system in order to diminish interference. The first is to stop transmission, or at least reduce its rate and power, when voice (or data) activity is absent or reduced. For a uniform population, this reduces the average signal power of all users and consequently the interference received by each user. The capacity is thus increased proportional to this overall rate reduction, provided the user population is large enough that the weak law of large numbers guarantees that the interference is nearly at its average value most of the time. We denote this factor as the voice activity gain, G_V. Numerous measurements on two-way telephone conversations [Brady, 1968] have established that voice is

---

[2] Conventional notation for noise density is N_0. Since, throughout this book, the density will be dominated by interference from other users, we employ the notation I_0, defined in (1.2), except when we deal with thermal background noise, where we shall use N_0. Throughout this chapter, thermal noise is assumed to be negligible. When it will be considered, N_0 will be taken as an additive component of the total interference density, I_0.

active only about three-eighths of the time, so that $G_V \approx 2.67$. Similarly, if we assume that the population of users is uniformly distributed in area over the single isolated cell, employing a sectored antenna reduces the interference and hence increases capacity by the antenna gain factor, G_A. Note that if the users are uniformly distributed in area, this is the classical definition of (two-dimensional) antenna gain [Lee, 1989], which is the received energy in the direction of the transmitter divided by the mean received energy, averaged over the circle. For a three-sectored antenna, this gain factor is less than 3. If we take the loss from ideal gain to be 1 dB, $G_A \approx 2.4$.

Finally, for a cellular system in which all users in all cells employ the common spectral allocation of W Hz, we must evaluate the interference introduced into each user's demodulator by all users in all other cells. A long geometrical and statistical argument,[3] based on the propagation losses of terrestrial transmission, leads to the conclusion that for a uniformly distributed population of users in all cells, each power-controlled by the appropriate base station, the total interference from users in all other cells equals approximately three-fifths of that caused by all users in the given cell. Thus, the capacity of Equation (1.4) must be reduced by the factor

$$f + 1 = \frac{\text{interference from other cells}}{\text{interference from given cell}} + 1 \approx 1.6.$$

Thus, introducing the voice and antenna gain factors, G_V and G_A, and the other-cell relative interference factor, f, into the capacity expression yields

$$k_u \approx \frac{W/R}{E_b/I_0} \frac{G_V G_A}{1 + f}. \tag{1.5}$$

With $G_V \approx 2.67$, $G_A \approx 2.4$, and $1 + f = 1.6$, this yields

$$k_u \approx \frac{4 \, W/R}{E_b/I_0}.$$

With well-designed modulation systems, employing error-correcting coding along with dual diversity antennas and a multipath diversity-combining (Rake) receiver at the base station, it is possible to achieve a

---

[3] To be presented in Chapter 6.

high level of performance (low error rates) with $E_b/I_0 \approx 6$ dB. For our example, the achieved multiple access system capacity per cell is approximately

$$k_u \approx W/R \text{ users.}$$

This will be shown to be nearly an order of magnitude greater than that achievable by systems that allocate disjoint resources to individual users, with commonly employed reuse factors.

This has been only a gross estimate, of course, intended to highlight the fundamental parameters in the determination of spread spectrum multiple access user capacity. Much more detailed and sophisticated analysis is needed to refine and validate this coarse approximation. That analysis is the purpose of the remainder of this text, culminating in a more complete and accurate analysis of capacity in Chapter 6.

1.4 Organization of the Text

The first issue to be dealt with in developing a spread spectrum system is the generation of a noiselike signal. This can be achieved in a variety of ways, but the generating process must be easily implemented and reproducible, because the same process must be generated at the transmitter for spreading and at the receiver for despreading, thus reconstituting the original information process. It will be shown that this "pseudonoise" spreading process is most easily implemented as a linear binary sequence generator followed by a linear filter. The necessary properties of the sequence generator and of the linear filter will be developed in Chapter 2. It is equally important in realizing a spread spectrum system to implement a synchronization technique that allows the receiver to synchronize the random signal that it generates to the signal received from the transmitter. Techniques both for initially acquiring such synchronization and for tracking it through variations due to motion and random effects will be treated in Chapter 3.

As illustrated by the fundamental capacity or interference–tolerance equation (1.5), the lower the E_b/I_0 ratio at which the communication system can perform properly, the higher its utilization efficiency. The modulation and its inherent forward error-correcting technique should be designed to allow the demodulation and associated error-correcting decoder at the receiver to operate at the lowest E_b/I_0 consistent with the required error rate and a practical level of complexity. This task is complicated by the fact that signal propagation is subject to multipath and, hence, fading characteristics. The spread spectrum nature of the signal provides ways of mitigating these degrading effects. If phase coherence can be maintained between transmitter and receiver, better performance is achieved. If not, noncoherent demodulation is required. Both cases will be considered, as well as means for achieving phase coherence. Such demodulation techniques are treated in Chapter 4, and their further amelioration, through coding and interleaving, is dealt with in Chapter 5. In many ways, these two chapters deal with digital modulation and coding systems that are more general than those employed in spread spectrum systems. However, the treatment will emphasize both the challenges and the advantages and simplifications that are possible when the signals are wideband and have noiselike characteristics. One important feature of spread spectrum multiple access systems is power control of each user by the base station with which it is communicating. The impact of power-control accuracy on performance is of particular interest and will be treated in these chapters.

Returning to broader system issues, in Chapter 6 the interference contribution of other-cell users and other-cell base stations will be evaluated, taking into account both the geometric dependence and the random nature of terrestrial signal propagation. These results, as well as those of Chapters 4 and 5 on modulation, coding, and power-control performance, will be used to analyze cell coverage and the true system capacity measured in erlangs per hertz of bandwidth per cell, for both reverse and forward links. Techniques for improving performance, including handoff between cells, and the use of sectored and distributed antennas and interference cancellation are also covered in Chapter 6.

CHAPTER TWO

Random and Pseudorandom Signal Generation

2.1 Purpose

Chapter 1 described the potential advantages of transmitting signals that appear noiselike and random. To be used in realizable systems, such signals must be constructed from a finite number of randomly preselected stored parameters. Equally important, the signals must be generated at the receiver as well and must be synchronized to coincide perfectly with the timing of the received transmission. This chapter deals with random signal generation, and the next chapter with the synchronization problem.

Because only a finite number of parameters can be stored at both transmitter and receiver locations, following the guidelines established by Nyquist's sampling theorem, the random waveform's numerical values need only be specified as samples at time intervals inversely proportional to the bandwidth occupied by the signals. Passing these samples through a linear filter generates the entire time-continuous waveform as an interpolation of the input samples. Strictly speaking, the signals should appear as Gaussian noise, which would dictate that each sample should approximate a Gaussian random variable. However, this would require specifying enough bits per sample to correspond to the quantization accuracy desired. We shall, instead, limit complexity by specifying only one bit per sample, corresponding to a binary sequence. Even with this drastic simplification, the effect of using such a random binary waveform is nearly the same as if Gaussian noise waveforms were used. The fact that binary random waveforms can be easily and flexibly modulated with an information-bearing digital signal has valuable practical consequences.

2.2 Pseudorandom Sequences

A binary independent random sequence, known in probability theory literature as a Bernoulli sequence, is sometimes referred to in engineering literature as a "coin-flipping" sequence: each "0" or "1" corresponds to a "heads" or "tails" outcome in a succession of independent coin-flip experiments. Even this simplified random sequence, however, would require arbitrarily large storage at both transmitter and receiver. We shall now demonstrate that the key "randomness" properties of a Bernoulli sequence can be successfully mimicked by a long deterministic periodic sequence that can be generated by a simple linear operation specified by a moderate number (tens) of binary parameters (bits). Thus, the only random variable is the starting point of the sequence.

Before we investigate the generation process for such "pseudorandom" sequences, it is important to specify key randomness properties that are to be achieved by the ultimately deterministic sequences. Traditionally, following Golomb [1967], these three properties are classified as R.1, R.2, and R.3, as follows:

R.1: Relative frequencies of "0" and "1" are each $\frac{1}{2}$.

R.2: Run lengths (of zeros or ones) are as expected in a coin-flipping experiment; half of all run lengths are unity; one-quarter are of length two; one-eighth are of length 3; a fraction $1/2^n$ of all runs are of length n for all finite n.

R.3: If the random sequence is shifted by any nonzero number of elements, the resulting sequence will have an equal number of agreements and disagreements with the original sequence.

A deterministically generated sequence that nearly satisfies (R.1) through (R.3), within extremely small discrepancies, will be referred to as a *pseudorandom sequence*. We shall provide a more precise definition after considering the generation process in the next section.

2.2.1 Maximal Length Linear Shift Register Sequences

We consider the deterministic linear binary sequence generator of Figure 2.1.

Each clock time the register shifts all contents to the right. The sequence $\{a_n, n$ any integer$\}$ propagates through with each term generated linearly

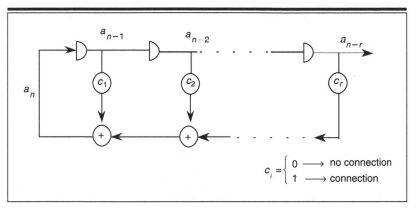

Figure 2.1 Linear shift register sequence generator.

from the preceding r terms according to the formula

$$a_n = c_1 a_{n-1} + c_2 a_{n-2} + \cdots + c_r a_{n-r} = \sum_{i=1}^{r} c_i a_{n-i}. \qquad (2.1)$$

Here, all terms are binary (0 or 1), c_1 to c_r are connection variables (1 for connection, 0 for no connection), and ordinary multiplication rules hold, but addition is modulo-2 (or "exclusive-OR"), meaning that $0 + 1 = 1 + 0 = 1$ but $0 + 0 = 1 + 1 = 0$. With these rules, all operations are linear[1] and the distributive law $a(b + c) = ab + ac$ applies. The sequence $\{a_n\}$ so generated may be doubly infinite. However, considering only terms with nonnegative index, we define the generating function of the sequence as

$$G(D) \triangleq a_0 + a_1 D + a_2 D^2 + \cdots = \sum_{n=0}^{\infty} a_n D^n, \qquad (2.2)$$

where D is the delay operator, and the power of D of each term of this polynomial corresponds to the number of units (clock cycles) of delay for that term.

---

[1] Alternatively, instead of zeros and ones, we will subsequently let the symbols be binary real numbers $+1$ and -1, keep the tap coefficients c_k as before, but replace all adders in the feedback paths by ordinary multipliers. The result is the same, but the operations no longer appear linear in the classical sense.

Combining (2.1) and (2.2), we may reduce the latter to the finite recurrence relation

$$
G(D) = \sum_{n=0}^{\infty} a_n D^n = \sum_{n=0}^{\infty} \sum_{i=1}^{r} c_i a_{n-i} D^n
$$

$$
= \sum_{i=1}^{r} c_i D^i \left[\sum_{n=0}^{\infty} a_{n-i} D^{n-i} \right]
$$

$$
= \sum_{i=1}^{r} c_i D^i [a_{-i} D^{-i} + \cdots + a_{-1} D^{-1} + G(D)].
$$

From this, $G(D)$ can be expressed as a ratio of finite polynomials because

$$
G(D) \left(1 - \sum_{i=1}^{r} c_i D^i \right) = \sum_{i=1}^{r} c_i D^i (a_{-i} D^{-i} + \cdots + a_{-1} D^{-1}),
$$

or

$$
G(D) = \frac{\displaystyle\sum_{i=1}^{r} c_i D^i (a_{-i} D^{-i} + \cdots + a_{-1} D^{-1})}{1 - \displaystyle\sum_{i=1}^{r} c_i D^i} \triangleq \frac{g_0(D)}{f(D)}, \qquad (2.3)
$$

where

$$
f(D) = 1 - \sum_{i=1}^{r} c_i D^i \qquad (2.4)
$$

is called the *characteristic polynomial* of the shift register sequence generator and depends solely on the connection vector $c_1, \ldots, c_r$. The polynomial $g_0(D)$ depends as well on the *initial condition vector* $a_{-r}, a_{-r-1}, \ldots, a_{-1}$, the contents of the register just before the a_0 term is generated. It may be written as

$$
g_0(D) = \sum_{i=1}^{r} c_i (a_{-i} + a_{-i+1} D + \cdots + a_{-1} D^{i-1})
$$

$$
\begin{aligned}
= \; & c_1 a_{-1} \\
& + c_2 (a_{-2} + a_{-1} D) \\
& + c_3 (a_{-3} + a_{-2} D + a_{-1} D^2) \\
& + \cdots \\
& + c_r (a_{-r} + a_{-r+1} D + \cdots + a_{-1} D^{r-1}).
\end{aligned} \qquad (2.5)
$$

Note that of all the connection variables, at least $c_r = 1$, for otherwise the shift register would no longer need to have r stages. Furthermore, for the time being we consider the initial vector

$$a_{-r} = 1, a_{-r+1} = \cdots = a_{-2} = a_{-1} = 0,$$

in which case (2.5) and (2.3) reduce to

$$g_0(D) = 1, \qquad G(D) = \frac{1}{f(D)}. \qquad (2.6)$$

Using (2.3) and initially the special case (2.6), we proceed to establish three basic properties of linear shift register (LSR) sequences, which we label (P-1) through (P-3).

P-1: Every LSR sequence is periodic with period

$$P \le 2^r - 1. \qquad (2.7)$$

Proof: There are exactly $2^r - 1$ distinct nonzero vectors that may be the contents of the register at any given time. If the contents are the all-zeros, then all future contents will be all-zeros, and thus the period is 1. For any other case there can be at most $2^r - 1$ clock cycles during which the nonzero content vectors are all distinct, for this is the maximum number of possibilities. When a particular vector repeats after $P \le 2^r - 1$ cycles, it will proceed to repeat all successive contents, since the future history depends only on this initial vector, as shown by (2.3). Thus, the sequence is periodic with period $P \le 2^r - 1$.

This leads us to define a *maximum length* (linear) *shift register* (MLSR) sequence as an LSR sequence whose period $P = 2^r - 1$ for all nonzero initial vectors. Further, we have

P-2: For all but degenerate cases, the period P of $G(D)$ is the smallest positive integer P for which $f(D)$ divides $1 - D^P$. Degenerate refers to the case where $g_0(D)$ and $f(D)$ have factors in common, as discussed when we introduce the general case.

Proof of Special Case: We first prove this for the initial vector generating function $g_0(D) = 1$, and subsequently generalize. We show that if $G(D)$ has period P, then $f(D)$ divides $1 - D^P$, and conversely that if

$f(D)$ divides $1 - D^P$, then $G(D)$ must have period P. Thus, if $G(D)$ has period P with $g_0(D) = 1$, it can be written as

$$
\frac{1}{f(D)} = G(D) = (a_0 + a_1D + \cdots + a_{P-1}D^{P-1})
$$
$$
+ D^P(a_0 + a_1D + \cdots + a_{P-1}D^{P-1})
$$
$$
+ D^{2P}(a_0 + a_1D + \cdots + a_{P-1}D^{P-1}) + \cdots
$$
$$
= \frac{a_0 + a_1D + \cdots + a_{P-1}D^{P-1}}{1-D^P}.
$$

Hence,

$$
\frac{1 - D^P}{f(D)} = a_0 + a_1D + \cdots + a_{P-1}D^{P-1},
$$

which proves the direct result.

For the converse, if $f(D)$ divides $1 - D^P$, then

$$
\frac{1 - D^P}{f(D)} = a_0 + a_1D + \cdots + a_{P-1}D^{P-1}.
$$

Thus, since $g_0(D) = 1$,

$$
G(D) = \frac{1}{f(D)} = \frac{a_0 + a_1D + \cdots + a_{P-1}D^{P-1}}{1 - D^P}
$$
$$
= (1 + D^P + D^{2P} + \cdots)(a_0 + a_1D + \cdots + a_{P-1}D^{P-1}),
$$

which implies that $G(D)$ is periodic with period P.

Proof of General Case: We now identify as degenerate the condition that $g_0(D)$ and $f(D)$ are not relatively prime. Then, according to (2.3), the degree of $f(D)$ is reduced, implying that the shift register length may be reduced. Thus, for nondegenerate cases—meaning that $g_0(D)$ and $f(D)$ have no common (polynomial) factors—we modify the direct proof for the special case. Note that if $G(D)$ has period P,

$$
\frac{g_0(D)}{f(D)} = G(D) = \frac{a_0 + a_1D + \cdots + a_{P-1}D^{P-1}}{1 - D^P},
$$

so that

$$g_0(D)(1 - D^P) = (a_0 + a_1D + \cdots + a_{P-1}D^{P-1})f(D).$$

Thus, by the unique factorization theorem, since $g_0(D)$ and $f(D)$ have no factors in common, $f(D)$ must divide $1 - D^P$.

For the converse, if $f(D)$ divides $1 - D^P$, then, as for the special case,

$$\frac{1}{f(D)} = (a_0 + a_1D + \cdots + a_{P-1}D^{P-1})(1 + D^P + D^{2P} + \cdots).$$

But now the initial vector $g_0(D) = g_0 + g_1D + \cdots + g_{r-1}D_{r-1}$, with $g_k = 1$ for at least one value of $k \neq 0$. Thus,

$$
\begin{aligned}
G(D) &= \frac{g_0(D)}{f(D)} \\
&= g_0(a_0 + a_1D + \cdots + a_{P-1}D^{P-1})(1 + D^P + D^{2P} + \cdots) \\
&\quad + g_1D(a_0 + a_1D + \cdots + a_{P-1}D^{P-1})(1 + D^P + D^{2P} + \cdots) \\
&\quad + \cdots \\
&\quad + g_{r-1}D^{r-1}(a_0 + a_1D + \cdots + a_{P-1}D^{P-1}) \\
&\quad \times (1 + D^P + D^{2P} + \cdots).
\end{aligned}
$$

Thus, each term of the sum is a periodic sequence with period P, and the sum of such periodic sequences is also periodic with period P.

This leads us to the third property, which deals with avoiding characteristic polynomials $f(D)$ that can be factored. For if $f(D)$ has a factor, then there will be some initial vector $g_0(D)$ that corresponds to a factor of $f(D)$. This causes a degeneracy as defined earlier, which reduces r and hence the period $P \leq 2^r - 1$.

Specifically,

P-3: A necessary condition for $G(D)$ to generate an MLSR sequence (with $P = 2^r - 1$) is that $f(D)$ of degree r be irreducible (not factorable).

Proof: Suppose $G(D)$ is an MLSR sequence with $P = 2^r - 1$, but $f(D)$ is factorable. We show that this leads to the inconsistent conclusion that $P < 2^r - 1$, and hence we prove the property by contradiction.

Without loss of generality, we may take the initial vector as 100, . . . , 0 [$g_0(D) = 1$]: If $G(D)$ is MLSR, the sequence has period $2^r - 1$, and consequently every nonzero vector will at some time be in the register and any one can be treated as the initial vector.

If $f(D)$ is factorable, there exist polynomials $s(D)$ and $t(D)$ of degrees $r_s \geq 1$ and $r_t \geq 1$ such that $f(D) = s(D)t(D)$, where $r = r_s + r_t$. Then, after a partial fraction expansion is performed, we obtain

$$G(D) = \frac{1}{f(D)} = \frac{\alpha(D)}{s(D)} + \frac{\beta(D)}{t(D)},$$

with

$$\text{period}\left[\frac{\alpha(D)}{s(D)}\right] \leq 2^{r_s} - 1, \qquad \text{period}\left[\frac{\beta(D)}{t(D)}\right] \leq 2^{r_t} - 1.$$

Hence,

$$\text{period}\,[G(D)] \leq \text{period}\left[\frac{\alpha(D)}{s(D)}\right] \cdot \text{period}\left[\frac{\beta(D)}{t(D)}\right]$$

$$\leq (2^{r_s} - 1)(2^{r_t} - 1) = 2^r - 2^{r_s} - 2^{r_t} + 1 \leq 2^r - 3,$$

which is a *contradiction*. Hence, $f(D)$ must be *irreducible*.

Unfortunately, though P-3 is a *necessary* condition, it is *not sufficient*. For a counterexample, take $r = 4$ with desired $P = 2^4 - 1 = 15$. Consider $f(D) = 1 + D + D^2 + D^3 + D^4$, which is irreducible. However, $f(D)$ divides $1 - D^5$ and hence has period 5 rather than 15. To obtain period 15, one may use instead the fourth-order irreducible polynomial $f(D) = 1 + D + D^4$, which divides $1 - D^{15}$ but no polynomial $1 - D^k$ for $k < 15$.

Such irreducible polynomials of degree r that generate an MLSR sequence of period $P = 2^r - 1$ are called *primitive*. Fortunately, primitive polynomials exist for all degrees $r > 1$. In fact, by use of algebraic theory well beyond the scope of our needs here, there is a very useful formula [Golomb, 1967] for the number of primitive polynomials of degree r:

$$N_P(r) = \frac{2^r - 1}{r} \prod_{i=1}^{J} \frac{P_i - 1}{P_i}. \qquad (2.8)$$

$\{P_i, i = 1, 2, \ldots, J\}$ is the prime decomposition of $2^r - 1$, i.e.,

$$2^r - 1 = \prod_{i=1}^{J} P_i^{e_i},$$

where e_i is an integer.

Thus, for example, for $r = 2$ through 6 we have, respectively,

$$r = 2, \quad 2^r - 1 = 3, \qquad N_P(2) = \frac{3}{2} \cdot \frac{2}{3} = 1;$$

$$r = 3, \quad 2^r - 1 = 7, \qquad N_P(3) = \frac{7}{3} \cdot \frac{6}{7} = 2;$$

$$r = 4, \quad 2^r - 1 = 15 = 5 \cdot 3, \qquad N_P(4) = \frac{15}{4} \cdot \frac{4}{5} \cdot \frac{2}{3} = 2;$$

$$r = 5, \quad 2^r - 1 = 31, \qquad N_P(5) = \frac{31}{5} \cdot \frac{30}{31} = 6;$$

$$r = 6, \quad 2^r - 1 = 63 = 7 \cdot 3^2, \qquad N_P(6) = \frac{63}{6} \cdot \frac{6}{7} \cdot \frac{2}{3} = 6.$$

Finding primitive polynomials becomes more of a chore as r becomes larger, but tables exist up to very large values of r [Peterson and Weldon, 1972; Simon et al, 1985]. Typical values of interest are for r between 10 and 50.

2.2.2 Randomness Properties of MLSR Sequences

We now proceed to demonstrate that MLSR sequences nearly satisfy the randomness properties of coin-flipping sequences as listed at the beginning of Section 2.2. We must recognize at the outset that the parameters of the generation process [the c_i coefficients of the generator polynomial $f(D)$] are deterministic. The only random parameters are the r terms of the initial vector or, equivalently, its time shift.

R.1: Balanced Property

Suppose we examine the first stage of the register of Figure 2.1 as the entire sequence is shifted through it. We may view this as the last (rightmost) bit of the r-dimensional contents vector at each clock cycle. Enumerating the contents vectors of the shift register as it generates an MLSR sequence is tantamount to enumerating all possible $2^r - 1$ binary vectors of length r, excluding the all-zeros. But if we included the all-zeros, we would have a precisely balanced sequence: Among all 2^r binary vectors, exactly half are even (having a zero in the last position) and half are odd (having a one there). Because we exclude the all-zeros, the balance is offset by 1 out of 2^r. Thus, of the $2^r - 1$ terms of the MLSR sequence, 2^{r-1} are ones and $2^{r-1} - 1$ are zeros. If we treat the initial vector as random, or equivalently assume a randomly selected starting point, the probabilities that at a particular clock cycle the shift register output is a zero or a one, respectively, are

$$\Pr(0) = \frac{2^{r-1} - 1}{2^r - 1} = \frac{1}{2}\left(1 - \frac{1}{P}\right),$$

$$\Pr(1) = \frac{2^{r-1}}{2^r - 1} = \frac{1}{2}\left(1 + \frac{1}{P}\right).$$

Thus, the unbalance is $1/P$. For $r = 10$, 30, and 50, $1/P$ is approximately 10^{-3}, 10^{-9}, and 10^{-15}, respectively.

R.2: Run Length Property

We proceed as for R.1, but now examine $n + 2$ successive stages of the register, where $n \le r - 2$. Consider all contents of the alternate forms

$$0x_1x_2 \cdots x_n0 \quad \text{and} \quad 1x_1x_2 \cdots x_n1.$$

Of the 2^n possibilities in each case, there will be just one that has run length exactly n: the all-ones of length n in the first case, and the all-zeros of length n in the second. Thus, of all the 2^{n+1} possibilities for the two cases, exactly two have run length n. Consequently, a fraction 2^{-n} have run length n for $1 \le n \le r - 2$. Now consider run lengths $r - 1$. For this to occur, the entire contents of the register at some point must be either

$$\underset{\underset{r-1}{\longleftarrow}}{00 \ \ldots \ 0} \ 1 \qquad \text{or} \qquad \underset{\underset{r-1}{\longleftarrow}}{11 \ \ldots \ 1} \ 0$$

In both cases, since all r stage contents are specified, there can be only one possible output. In the first case, it must be a one. Otherwise, we would enter the all-zeros state, which is excluded. In the second case, it must also be a one. This is the only way that the all-ones vector can occur, which is one of the $2^r - 1$ state vectors that must occur in the sequence. (Note that for this second case, the next one must be followed by a zero; otherwise, the all-ones state would reoccur indefinitely.) Thus, there is only one possible run of length $r - 1$ (the zeros) and one possible run of length r (the ones), each occurring with relative frequency $1/2^{(r-1)}$.

We conclude therefore that the relative frequency of run length n (zeros or ones) is $1/2^n$ for all $n \le r - 1$ and $1/2^{(r-1)}$ for $n = r$, with no run lengths possible for $n > r$. Note, as a check, that the sum of the relative frequencies is unity.

R.3: Delay and Add Property

Consider any two clock cycle shifts of an MLSR sequence. If we take the entire sequence of length $P = 2^r - 1$ and shift it by an arbitrary number of cycles $\tau < P$, we obtain the MLSR sequence for a different initial vector. Now, if we add the original and shifted sequences, term by term modulo-2, we obtain a new sequence that is itself the MLSR sequence with yet another starting vector, and hence another initial point. Symbolically, using the polynomial notation of (2.2), we denote the original sequence by $G_0(D)$, the shifted sequence by $G_\tau(D)$, both of length $2^r - 1$, and their respective initial conditions $g_0(D)$ and $g_\tau(D)$, as defined by (2.5). Now since $G_0(D)$ and $G_\tau(D)$ are generated linearly from $g_0(D)$ and $g_\tau(D)$ according to (2.3),

$$G_0(D) = \frac{g_0(D)}{f(D)}, \qquad G_\tau(D) = \frac{g_\tau(D)}{f(D)},$$

where $f(D)$ is the characteristic (primitive) polynomial of the MLSR. Further, since the polynomial operations are linear, it follows from the distributive law that the modulo-2 sum of the two MLSR sequences has polynomial

$$G_0(D) + G_\tau(D) = \frac{g_0(D) + g_\tau(D)}{f(D)}.$$

This means that it can be generated by the initial condition polynomial $g_0(D) + g_\tau(D)$ (whose components are the modulo-2 sum of the respective components), but this is itself another valid initial vector. Hence, the sequence so generated, whose generator polynomial is $G_0(D) + G_\tau(D)$, is itself a time shift of the same MLSR sequence. This is the *delay-and-add property.*

Using this and property R.1, it follows that the two sequences $G_0(D)$ and $G_\tau(D)$, each of length $2^r - 1$, differ in 2^{r-1} positions and agree in $2^{r-1} - 1$. This is because their modulo-2 sum has a one for each disagreement and a zero for each agreement and is itself an MLSR sequence, which must therefore have the quasi-balanced property R.1.

R.1 and R.3 can also be stated in terms of time averages and correlations based on mapping each "0" to a $+1$ real number value and each "1" to a -1 real number value. In these terms we have properties

$$(\textbf{R.1})': \frac{1}{P} \sum_{n=1}^{P} \alpha_n = -\frac{1}{P}, \qquad (2.9)$$

where α_n is the real value equivalent of the nth term of the MLSR sequence, a_n. (2.9) follows from the quasi-balanced property; it would be zero for perfect balance.

Similarly, for $\tau \neq 0$,

$$(\textbf{R.3})': \frac{1}{P} \sum_{n=1}^{P} \alpha_n \cdot \alpha_{n+\tau} = -\frac{1}{P}. \qquad (2.10)$$

Here, as previously noted, real multiplication takes the place of modulo-2 addition with the same rules $(+ \cdot - = - \cdot + = -; + \cdot + = - \cdot - = +)$.

(2.9) is the temporal mean and (2.10) is the temporal correlation. However, properties (R.1)′ and (R.3)′ can also be taken as ensemble averages rather than time averages. Note that the (deterministic) MLSR sequence becomes a stationary ergodic sequence if we treat the initial vector (or time of observation) as a uniformly distributed random vector (or time variable).

2.2.3 Conclusion

Based on the original exposition of Golomb [1967], we have thus shown that with the slight unbalance of $1/P$ (less than one part in a million for $r > 20$), an MLSR sequence is indistinguishable from a Bernoulli or "coin-flipping" binary sequence, at least with respect to randomness properties (R.1) through (R.3), as long as the initial vector or time is chosen randomly. In all that follows, we shall treat MLSR and coin-flipping sequences interchangeably.

2.3 Generating Pseudorandom Signals (Pseudonoise) from Pseudorandom Sequences

At the end of the last section, we suggested mapping from binary logical symbols, 0 and 1, to real values $+1$ and -1. We now combine this with physical waveforms, creating the rudimentary spread spectrum digital communication system shown in Figure 2.2. Random binary numbers are turned into a noiselike waveform by modulating a periodic impulse stream of period T_c, which is the clock period of the (pseudo) random sequence generator. T_c is also called the *chip duration,* where chip refers to the signal corresponding to an individual term of the random sequence. The signal energy for each chip is designated E_c. The sign is established (modulated) by the binary random sequence value ($+1$ or -1) multiplied by the value of the binary data during the period T_c (with 0 or 1 also mapped to $+1$ or -1). The data sequence (bit or symbol) duration is typically a multiple of the chip duration, so the chip sequence value $x_n(k)$ will typically remain constant over several chips (values of n). For now, we will take the data sequence to be uncoded. However, this model also applies to a system employing error-correcting coding redundancy, in which case $x_n(k)$ is the value (± 1) during a chip duration T_c (which remains constant over all the chips within the code symbol duration). The index k is used to indicate that the modulator and signal generator pertain to the kth user of the multiple access system.

The impulse modulator is, of course, a fictitious construction. It provides a convenient linear model of the waveform generating process. In Figure 2.3, we examine the second-order statistics of the impulse train randomly modulated by $+1$ and -1 values of the product of the data and random sequence. Impulses (Dirac delta functions) are nonphysical and represent the limit of a very narrow, very tall pulse, as the pulse width Δ approaches zero and its amplitude $\sqrt{E_c}/\Delta$ approaches infinity. We show the nonzero Δ case in Figure 2.3b. From this and properties R-1 and R-3 of the last section, we obtain the correlation function $R(\tau)$ for the nonzero Δ case in Figure 2.3c. Finally, letting $\Delta \to 0$, we obtain the correlation function of the impulse train in Figure 2.3d. This is itself a single impulse or Dirac delta function of area E_c/T_c at $\tau = 0$, which is equivalent to *white noise* of two-sided density E_c/T_c watts/Hz.[2] Hence, the term *pseudonoise*

---

[2] In order to justify that the process is stationary, we must take the initial time to be a random variable, uniformly distributed over the interval $(0, T_c)$.

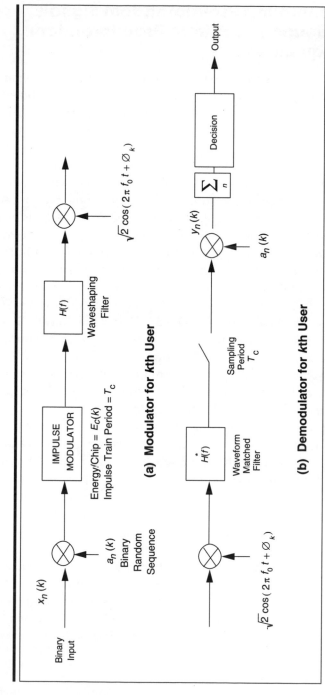

Figure 2.2 BPSK spread spectrum modulator–demodulator. (a) Modulator for *k*th user. (b) Demodulator for *k*th user.

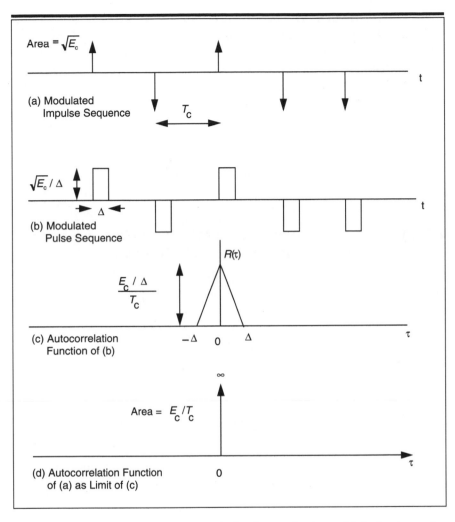

Figure 2.3 Correlation function computation for randomly modulated pulse and impulse sequences.

is used to describe a random signal generated as in Figure 2.2. The linear filter with transfer function $H(f)$ and impulse response $h(t)$ serves to contain the signal in the desired spectral allocation. The last function in the transmitter is upconversion to the desired center frequency.

In the presence of additive white Gaussian noise, the receiver performs the matched operations shown in Figure 2.2b. These include downconversion, matched filtering with the complex conjugate of the transmit filter $H^*(f)$, and sampling at the chip intervals. The matched filter maximizes output signal-to-noise ratio only when the interference is additive white Gaussian noise. The background noise can be taken to be white Gaussian. However, the interference caused by other users sharing the band of user k is not white, since they are all assumed to use the same spectral shaping filter. On the other hand, there will also be yet other users in contiguous bands, so the composite of all other users' signals in all bands may as well be taken to be essentially Gaussian noise with uniform spectral density, thus justifying the matched filter approach. Furthermore, as shown in Appendix 2A, when the shaping filter $H(f)$ is strictly band-limited, the optimum receiver filter is itself bandlimited, matched to $H(f)$, and both are constant over the bandlimited spectrum.

2.3.1 First- and Second-Order Statistics of Demodulator Output in Multiple Access Interference

Referring to Figure 2.2, we now examine the demodulator output per chip, $y_n(k)$, of the kth user at the nth chip sampling time, given that the binary data input for this chip is $x_n(k)$, which is $+1$ or -1. We assume, initially, that the pseudorandom sequence generator at the receiver is per-fectly synchronized with the received signal. As with any digital commu-nication system, spread spectrum or not, there are four components of the demodulator output:

(a) the desired output, which depends only on $x_n(k)$;

(b) the interchip interference components, which depend only on $x_{n+m}(k)$, $m \neq 0$ (usually called intersymbol interference [ISI] for nonspread digital demodulation);

(c) the component due to background noise, which shall be assumed to be white of one-sided density N_0 watts/Hz;

(d) the other-user interference components, which depend on $x_{n+m}(j)$ for all $j \neq k$ and all m.

All but (c) depend also on the propagation characteristics of the signals' transmission. For this introductory treatment, we shall assume that all signal powers are fixed and known, generalizing later in Chapter 4 to multipath and fading channels.

Since all the interference components have zero mean, the mean of $y_n(k)$ is a function of $x_n(k)$. Hence, using Parseval's theorem and the fact that $a_n^2(k) = 1$, we have

$$E[y_n(k) \mid x_n(k)] = \sqrt{E_c(k)}x_n(k) \int_{-\infty}^{\infty} \mid H(f) \mid^2 df. \qquad (2.11)$$

$E_c(k)$ is the chip energy for the kth user, which requires that the filter gain be normalized, so that

$$\int_{-\infty}^{\infty} \mid H(f) \mid^2 df = 1. \qquad (2.12)$$

The interchip components depend on $a_{n+m}(k) \; x_{n+m}(k)$, $m \neq 0$, which are also binary random variables, independent of $x_n(k)$ because of the randomness properties established in Section 2.2. If we treated these as deterministic values, their effect on the nth sample at the demodulator would be

$$\sqrt{E_c(k)}a_n(k) \sum_{m \neq 0} a_{n+m}(k)x_{n+m}(k) \int_{-\infty}^{\infty} \cos(2\pi \, mfT_c) \mid H(f) \mid^2 df.$$

This follows from the fact that the Fourier transform of any impulse delayed by m chips, filtered by $H(f)$ and $H^*(f)$, is $\exp(-2\pi \, imfT_c) \mid H(f) \mid^2$, and $\mid H(f) \mid^2$ is an even function. However, since the $a_{n+m}(k)$ are all independent binary random variables with zero means and unit variances, it follows that the interchip interference components contribute nothing to the mean of the output, but add to its variance the quantity

$$V_I = \mathrm{Var}_I[y_n(k) \mid x_n(k)]$$
$$= E_c(k) \sum_{m \neq 0} \left[\int_{-\infty}^{\infty} \cos(2\pi \, mfT_c) \mid H(f) \mid^2 df \right]^2, \qquad (2.13)$$

where the subscript I refers to interchip interference.

We observe further that the other two components, background white noise and other-user interference, are independent of $x_n(k)$ for all n. Thus, the variance due to background noise is just the effect of white noise of one-sided density N_0 on the received filter whose transfer function is $H^*(f)$, which must therefore contribute

$$V_N = \mathrm{Var}_N[y_n(k) \mid x_n(k)] = (N_0/2) \int_{-\infty}^{\infty} \mid H(f) \mid^2 df = N_0/2. \qquad (2.14)$$

The subscript N refers to background noise.

Finally, any other user j has constant chip energy $E_c(j)$ modulated by binary variables $a_n(j)\, x_n(j)$ which are independent of and generally unsynchronized with those of user k.[3] In addition, the carrier phase of the jth user's modulator ϕ_j will differ from that of the kth user. Hence, the effect of the jth user's signal on the kth user's demodulator will be that of white noise with two-sided density $E_c(j)/T_c$ (see Figure 2.3d) passing through the tandem combination of two filters with combined transfer function $|H(f)|^2$. Taking into account the relative phases of the jth and kth users, which we momentarily take as fixed and given, the variance of the output of the kth demodulator due to the jth user's signal is

$$V_j = \mathrm{Var}_j[y_n(k) \mid x_n(k),\, \phi_j,\, \phi_k]$$

$$= [E_c(j)/T_c]\cos^2(\phi_j - \phi_k) \int_{-\infty}^{\infty} |H(f)|^4\, df. \qquad (2.15)$$

Here, as is normal, terms involving double the carrier frequency are assumed to be totally filtered out. Now because the relative phases of all users are random variables that are uniformly distributed between 0 and 2π, and all signals are mutually independent, averaged over all phases, the variance of the nth chip of the kth user due to all other user signals is

$$V_O = \sum_{j \neq k} E_{\phi_j}(V_j) = \sum_{j \neq k} E_c(j) \int_{-\infty}^{\infty} |H(f)|^4\, df/(2T_c). \qquad (2.16)$$

Consequently, since all random outputs contributing to the variance are independent, the first- and second-order statistics of the output of the nth chip of the kth user, as obtained from (2.11), (2.12), (2.13), (2.14), and (2.16), are

$$E[y_n(k) \mid x_n(k)] = \sqrt{E_c(k)}\, x_n(k),$$

$$\mathrm{Var}[y_n(k) \mid x_n(k)] = V_I + V_N + V_O, \qquad (2.17)$$

where

$$V_I = E_c(k) \sum_{m \neq 0} \left[\int_{-\infty}^{\infty} \cos(2\pi\, mfT_c)\, |H(f)|^2\, df \right]^2,$$

$$V_N = N_0/2, \qquad (2.18)$$

$$V_O = \sum_{j \neq k} E_c(j) \int_{-\infty}^{\infty} |H(f)|^4\, df/(2T_c).$$

---

[3] The relative timing of the impulse trains of different users is a random variable uniformly distributed over the interval $(0, T_c)$.

One reservation about these ensemble averages: if the number of users is small, phases might sometimes align themselves so that averaging on phase is not appropriate. We now show that (2.18) holds, without averaging on phase, if the carriers are quadriphase (QPSK) modulated by two pseudorandom sequences.

2.3.2 Statistics for QPSK Modulation by Pseudorandom Sequences

To avoid any dependence on carrier phase, we consider the QPSK spreading modulator and corresponding demodulator, shown in Figure 2.4. Note that the input data $x_n(k)$ are still biphase modulated. We assume that the receiver is tracking the transmitted carrier perfectly, and that the in-phase (I) and quadrature (Q) pseudorandom sequences are independent (or their generators are the same, but their initial vectors are separated by a randomly chosen offset). The means and variances of the in-phase and quadrature components are obtained as follows:

$$E[y_n^{(I)}(k) \mid x_n(k)] = E[y_n^{(Q)}(k) \mid x_n(k)] = x_n(k)\sqrt{E_c(k)}/2,$$

$$\mathrm{Var}_I[y_n^{(I)}(k)] = \mathrm{Var}_I[y_n^{(Q)}(k)] = V_I/4,$$

$$\mathrm{Var}_N[y_n^{(I)}(k)] = \mathrm{Var}_N[y_n^{(Q)}(k)] = N_0/4 = V_N/2,$$

$$V_O[y_n^{(I)}(k)/\phi_j, \phi_k] = \sum_{j \neq k} [E_c(j)/(4T_c)][\cos(\phi_j - \phi_k) + \sin(\phi_j - \phi_k)]^2$$

$$\times \int \mid H(f) \mid^4 df,$$

$$V_O[y_n^{(Q)}(k)/\phi_j, \phi_k] = \sum_{j \neq k} [E_c(j)/(4T_c)][\cos(\phi_j - \phi_k) - \sin(\phi_j - \phi_k)]^2$$

$$\times \int \mid H(f) \mid^4 df.$$

Finally, since $a_n^{(I)}(k)$ and $a_n^{(Q)}(k)$ are independent, $y_n^{(I)}(k)$ and $y_n^{(Q)}(k)$ are uncorrelated. Then, taking the signal processor in Figure 2.4b to perform in this case a summation of the in-phase and quadrature components, we obtain

$$y_n(k) = y_n^{(I)}(k) + y_n^{(Q)}k.$$

The mean and variance of $y_n(k)$ are the sums of the means and variances of the components, as follows:

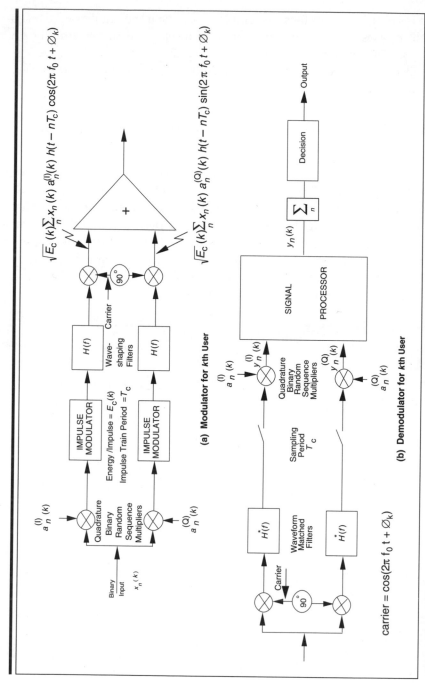

Figure 2.4 QPSK spread spectrum modulator–demodulator. (a) Modulator for *k*th user. (b) Demodulator for *k*th user.

$$E[y_n(k) \mid x_n(k)] = \sqrt{E_c(k)}x_n(k),$$

$$\text{Var}[y_n(k) \mid x_n(k)] = \left(\frac{1}{2}\right)V_1 + V_N + V_O, \qquad (2.19)$$

where V_I, V_N, and V_O are given by (2.18).

We conclude then that the first- and second-order statistics for QPSK modulation by pseudorandom sequence are exactly the same as those for BPSK modulation, except that the interchip interference V_I is reduced in half (through the use of independent spreading sequences on the I and Q channels). On the other hand, V_O, the variance of other-user noise, is independent of the relative phase, so that it is not necessary to take the ensemble average as was done in (2.16) to obtain the result of (2.18).

Also, if the phase of the kth user's demodulator ϕ_k' is not equal to that of the modulator ϕ_k, the means for both the BPSK and QPSK cases, as determined from Figures 2.2 and 2.4, are reduced to

$$E[y_n(k) \mid x_n(k)] = \sqrt{E_c(k)}x_n(k)\cos(\phi_k - \phi_k'). \qquad (2.20)$$

The background noise variance, V_N, and the other-user interference variance, V_O, are obviously unaffected by any phase inaccuracy in the demodulator.

2.3.3 Examples

It is clear from (2.18) that the variance depends heavily on the form of the linear wave-shaping filter transfer function. We now evaluate this for the two limiting cases of time-limited and bandlimited filters. In the first case, we take the impulse response to be a constant rectangular pulse of duration T_c. Thus, the waveform consists of a succession of pulses, each of duration T_c and sign determined by the pseudorandom sequence $x_n(k)$. Thus, normalizing according to the condition (2.12), we have

$$h(t) = (1/\sqrt{T_c})[u(t) - u(t - T_c)], \qquad (2.21)$$

where $u(\ )$ is the unit step function. Consequently, the constant *time-limited* transfer function is

$$H(f) = \int_0^{T_c} h(t)\exp(-2\pi \, ift)\,dt$$

$$= \sqrt{T_c}\exp(-i\pi fT_c)\sin(\pi fT_c)/(\pi fT_c). \qquad (2.22)$$

At the other extreme, we have a constant *bandlimited* transfer function

$$H(f) = \frac{1}{\sqrt{W}} \left[u\left(f + \frac{W}{2}\right) - u\left(f - \frac{W}{2}\right) \right], \qquad (2.23)$$

for which

$$h(t) = \int_{-W/2}^{W/2} H(f) \exp(2\pi ift)\, df = \sqrt{W} \sin(\pi Wt)/(\pi Wt),$$

$$\text{where } W \triangleq 1/T_c. \quad (2.24)$$

In both cases, the interchip interference is exactly zero, since $h(nT_c) = 0$ for all $n \neq 0$.

For the *constant h(t) time-limited* pulse case, the other-user noise integral can be obtained from Parseval's theorem as

$$\int |H(f)|^4 \, df = \int |R(t)|^2 \, dt,$$

where $R(t)$ is the inverse Fourier transform of $|H(f)|^2$. Thus,

$$R(t) = \int_{-\infty}^{\infty} T_c \left(\frac{\sin \pi fT_c}{\pi fT_c}\right)^2 \exp(2\pi ift)\, df$$

$$= \begin{cases} 1 - |t|/T_c, & |t| < T_c, \\ 0 & \text{otherwise,} \end{cases} \qquad (2.25)$$

from which it follows that

$$\frac{1}{T_c} \int_{-\infty}^{\infty} |H(f)|^4 \, df = \frac{2}{T_c} \int_0^{T_c} (1 - t/T_c)^2 \, dt = \frac{2}{3}$$

$$\text{(time-limited).} \quad (2.26)$$

This result was first obtained by an alternate derivation [Pursley, 1977].

For the *constant H(f) bandlimited* case, we have trivially,

$$\frac{1}{T_c} \int_{-\infty}^{\infty} |H(f)|^4 \, df = \frac{1}{W} \int_{-W/2}^{W/2} df = 1 \qquad \text{(bandlimited).} \quad (2.27)$$

Note that in this case, since $|H(f)|^2 = H(f)/\sqrt{W}$, $R(t) = h(t)/\sqrt{W}$, as given by (2.24). Thus, in both cases, $R(nT_c) = 0$ for all $n \neq 0$. Thus, successive chip outputs are uncorrelated.

2.3.4 Bound for Bandlimited Spectrum

Suppose the spectrum is strictly bandlimited to $(-W/2, W/2)$, but is not necessarily constant. As always, we normalize the transfer function gain according to (2.12). It is of interest to determine the minimum value of the variance. The other-user component of the variance can be lower-bounded by using the Schwarz inequality:

$$\int_{-W/2}^{W/2} A^2(f)\, df \int_{-W/2}^{W/2} B^2(f)\, df \geq \left[\int_{-W/2}^{W/2} A(f)B(f)\, df \right]^2.$$

Letting $A(f) = |H(f)|^2$ and $B(f) = 1$, it follows that

$$\int_{-W/2}^{W/2} |H(f)|^4\, df \cdot W \geq \left[\int_{-W/2}^{W/2} |H(f)|^2\, df \right]^2 = 1.$$

Thus,

$$\int_{-W/2}^{W/2} |H(f)|^4\, df \geq \frac{1}{W}, \tag{2.28}$$

with equality if and only if

$$|H(f)|^2 = \frac{1}{W}, \qquad \frac{-W}{2} \leq f \leq \frac{W}{2}. \tag{2.28a}$$

But then, for constant $|H(f)|$, we have from (2.24) that $W = 1/Tc$. Thus,

$$\int |H(f)|^4\, df/T_c = 1.$$

Furthermore, for constant bandlimited $H(f)$, it follows from (2.24) that $h(nT_c) = 0$ for all $n \neq 0$, so that the interchip interference variance $V_I = 0$ and hence is also minimized. We conclude that, for a *bandlimited* spectrum, the lower bound on variance is

$$\mathrm{Var}[y_n(k)] \geq \left[N_0 + \sum_{j \neq k} E_c(j) \right] / 2, \tag{2.29}$$

and that this lower bound is achieved by the constant transfer function bandlimited filter of (2.23). Appendix 2A shows another derivation of the optimality of the constant bandlimited $H(f)$ *without* the a priori assumption that the receiving filter is matched to $H(f)$.

2.4 Error Probability for BPSK or QPSK with Constant Signals in Additive Gaussian Noise and Interference

Throughout this chapter we have assumed that all signals were at constant power throughout the transmission period. We shall reexamine this assumption in Chapter 4 when multipath and fading effects are considered. We computed second-order statistics, but did not examine their distributions. When the chip period is T_c and the bit rate is R, the number of chips per bit is $1/(RT_c)$.[4] Then the final processor in the demodulator must accumulate over as many chips as the input data $x_n(k)$ is constant, thus forming

$$Y = \sum_{n=1}^{1/(RT_c)} y_n(k).$$

In an uncoded BPSK system, the sign of this measurement is used to decide whether the bit was 0 or 1 ($+1$ or -1).
Then

$$E(Y) = \sum_{n=1}^{1/(RT_c)} E[y_n(k) \mid x_n(k)]$$

$$= \sqrt{E_c(k)} \sum_{n=1}^{1/(RT_c)} x_n(k) = \pm[1/(RT_c)]\sqrt{E_c(k)}. \tag{2.30}$$

Here, the sign depends on the sign of $x_n(k)$, which is constant over all chips n within the bit period, $1/R$.

Since the noise components of the individual chip outputs are essentially uncorrelated,

$$\mathrm{Var}(Y) = \sum_{n=1}^{1/(RT_c)} \mathrm{Var}[y_n(k) \mid x_n(k)]$$

$$= [1/(RT_c)](V_I + V_N + V_O). \tag{2.31}$$

Now if the number of chips per bit is large so that $1/(RT_c) \gg 1$, according to the central limit theorem, the variable Y will be nearly Gaussian.
Then the bit error probability is given by

---

[4] We shall generally assume that the system is implemented so that the symbol period is an integral number of chip times, so that $1/(RT_c)$ is an integer.

$$P_b = Q\left(\sqrt{\frac{[E(Y)]^2}{\mathrm{Var}\, Y}}\right)$$

$$= Q\left(\sqrt{\frac{E_c(k)/(RT_c)}{V_I + V_N + V_O}}\right) \qquad (2.32)$$

where

$$Q(z) \triangleq \int_z^\infty e^{-x^2/2}\, dx/\sqrt{2\pi}.$$

Taking the interchip interference to be negligible or zero, we have using (2.18)

$$P_b = Q\left(\sqrt{\frac{2E_c(k)/(RT_c)}{N_0 + \sum_{j\neq k} E_c(j)\int_{-\infty}^\infty |H(f)|^4\, df/T_c}}\right). \qquad (2.33)$$

The numerator is just twice the bit energy, E_b, and the denominator is the effective interference density.

Thus,

$$P_b = Q\left(\sqrt{\frac{2E_b}{I_0}}\right), \qquad (2.34)$$

where

$$\frac{E_b}{I_0} = \frac{E_c(k)/(RT_c)}{N_0 + \sum_{j\neq k} E_c(j)\int |H(f)|^4\, df/T_c}. \qquad (2.35)$$

(2.34) is, of course, the classical expression for bit error probability of an uncoded coherent BPSK or QPSK demodulator in additive white noise of density I_0.

Suppose we ignore background noise ($N_0 = 0$), as we did in Section 1.3, and assume that because of power control, all users are received by the base station receiver at the same power or chip energy level E_c. Then for a given E_b/I_0 level, determined from (2.34) for the required bit error probability, the total number of users, k_u, is obtained from (2.35) to be

$$k_u - 1 = \frac{1/(RT_c)}{\int \mid H(f) \mid^4 df/T_c} \cdot \frac{1}{E_b/I_0} . \qquad (2.36)$$

For a bandlimited spectrum, $W = 1/T_c$ and the integral in the denominator is lower-bounded by unity. That leaves us with

$$k_u - 1 \leq \frac{W/R}{E_b/I_0} , \qquad (2.37)$$

with equality for a constant bandlimited transfer function waveshaping filter. This is precisely Equation (1.4) of Section 1.3 and validates the intuitive results obtained there by dimensional analysis.

Optimum Receiver Filter for Bandlimited Spectrum

We show that with the bandlimited condition

$$| H(f) | = 0 \qquad \text{for } | f | > W/2, \qquad (2A.1)$$

the constant bandlimited filter of (2.28a) maximizes the output signal-to-interference ratio *without* the a priori assumption that the receiver filter is *matched* to the transmitter shaping filter.

We assume as before a normalized shaping filter

$$\int_{-W/2}^{W/2} | H(f) |^2 \, df = 1, \qquad (2A.2)$$

and we define the arbitrary receiving filter transfer function $G(f)$. Subject to the normalizing condition,

$$\left| \int_{-W/2}^{W/2} H(f)G(f) \, df \right|^2 = k_0(\text{constant}), \qquad (2A.3)$$

we seek to minimize the variance

$$\text{Var}(y_n) = (1/2)$$
$$\left\{ \sum_{j \neq k} [E_c(j)/T_c] \int_{-W/2}^{W/2} | H(f)G(f) |^2 \, df \right.$$
$$\left. + N_0 \int_{-W/2}^{W/2} | G(f) |^2 \, df \right\}. \qquad (2A.4)$$

This variance includes other-user interference and the background noise, but not interchip interference. The last is generally negligible and will turn out to be zero for the optimum choice.

Using the Schwarz inequality in two different ways and normalizing conditions (2A.2) and (2A.3), the two components of (2A.4) can be bounded as follows:

$$k_0 = \left| \int_{-W/2}^{W/2} H(f)G(f) \cdot 1 \, df \right|^2 \leq \int_{-W/2}^{W/2} | H(f)G(f) |^2 \, df \cdot W, \quad (2A.5)$$

and

$$k_0 = \left| \int_{-W/2}^{W/2} H(f)G(f) \, df \right|^2$$

$$\leq \int_{-W/2}^{W/2} | H(f) |^2 \, df \int_{-W/2}^{W/2} | G(f) |^2 \, df \quad\quad (2A.6)$$

$$= \int_{-W/2}^{W/2} | G(f) |^2 \, df.$$

It follows from the (2A.5) and (2A.6), letting $W = 1/T_c$, that the variance of (2A.4) is lower-bounded by

$$\text{Var}(y_n) \geq (k_0/2) \left\{ \sum_{j \neq k} E_c(j) + N_0 \right\}. \quad\quad (2A.7)$$

Without loss of generality, other than a scaling of the receiver filter gain, we may let $k_0 = 1$, which makes (2A.7) the same as (2.29). Note then that the optimizing filter must satisfy the conditions (2A.5) and (2A.6) as *equalities*. The second condition requires

$$G(f) = k'H^*(f), \quad\quad | f | \leq W/2, \quad\quad (2A.8)$$

where k' is a constant. This means that the receiver filter must be *matched* to the transmitter filter. The first condition requires

$$H(f)G(f) = k'', \quad\quad | f | \leq W/2, \quad\quad (2A.9)$$

where k'' is a constant. The signal-to-noise ratio is just the ratio of (2A.3) to (2A.4). Then it follows from (2A.8), (2A.9), and (2A.2) that the signal-to-noise ratio is maximized by the choice

$$| H(f) |^2 = 1/W. \quad\quad (2A.10)$$

Thus, both the transmitter shaping filter $H(f)$ and the receiver filter $G(f)$ are constant over the bandlimited spectrum $(-W/2, W/2)$. This is the same as (2.28a), but obtained without the a priori assumption that $G(f)$ is matched to $H(f)$.

CHAPTER THREE

Synchronization of Pseudorandom Signals

3.1 Purpose

The performance analysis of the spread spectrum modulation system presented in the previous chapter requires the receiver carrier's phase and frequency and its chip timing to be perfectly synchronized to that of the transmitted signal, as received with the appropriate propagation delay. This chapter considers methods for achieving this synchronization, particularly of the timing of the pseudorandom signals. It also covers the evaluation of the performance loss as a function of inaccuracies in receiver timing, as well as phase and frequency.

Provided the initial frequency error is relatively small, a number of practical considerations justify acquiring timing before acquiring accurate phase and frequency. The timing acquisition system and process are considered next. Tracking of timing is treated in later sections. After timing has been established, acquisition and tracking of phase and frequency are performed in the same way as for any digital communication system. This is discussed briefly in the final section.

3.2 Acquisition of Pseudorandom Signal Timing

Any acquisition process involves testing all likely hypotheses for the correct value of the parameter. Because of resource limitations, this testing is generally performed serially: each incorrect hypothesis is eliminated before the next is tested. For a small number of hypotheses, testing can be performed in parallel. In general, a combination of the two may be implemented, using all available parallel devices and assigning each a fraction of the total hypotheses. Another strategy will also be considered: performing two passes through the hypotheses, eliminating the least likely on the first pass, and revisiting each remaining hypothesis for a longer time on the second pass. This can be generalized to more passes, as will be discussed. Whichever search strategy is employed, the basic hypothesis testing device is the same. In the next subsection, this will be considered for a BPSK spreading waveform, and in the following subsection, for a QPSK spreading waveform.

3.2.1 Hypothesis Testing for BPSK Spreading

Time acquisition is the process of determining the time of the matched filter output peak for the chip corresponding to the first output (produced by the initial vector) of the pseudorandom linear sequence generator. This is required to synchronize the locally generated spreading sequence to that of the desired received signal. Strictly speaking, determining the exact sampling time is an estimation problem. However, to reduce the problem to one of finite dimensionality, we only test the hypothesis that a particular chip is the first in the sequence by sampling the matched filter once, or at most a few times, per chip. As will presently be shown, this results in performance degradation that increases monotonically with the time difference between the nearest tested sampling time and the point of exact synchronization. Once this approximate synchronization has been achieved, by the search process described in Section 3.4, the estimate is refined to the required accuracy by a closed-loop tracking technique, to be described in Section 3.5.

The basic hypothesis testing device is shown in Figure 3.1a. The front end for BPSK spreading is the same as that of the QPSK demodulator (Figure 2.4b). We shall justify the remaining functions of the signal processor, both linear and quadratic, in Section 3.3 and Appendix 3A. It suffices at this point to emphasize that the spreading sequence timing is determined before any phase measurement or tracking is attempted. Hence, the hypothesis testing device must be noncoherent, effectively measuring energy as shown in Figure 3.1a. Even so, if the frequency error is greater than the inverse of the testing time of each hypothesis, NT_c, it will be shown in Section 3.2.3 that the energy measurement is seriously degraded. Consequently, the hypothesis test becomes unreliable. To prevent this, it may be necessary to perform "postdetection integration" by accumulating $L > 1$ successive energy measurements, performed on the same time hypothesis, before making the decision, as shown by the last block of Figure 3.1a. These qualitative observations will be supported by the quantitative results of Section 3.3. But first, in this section, we establish

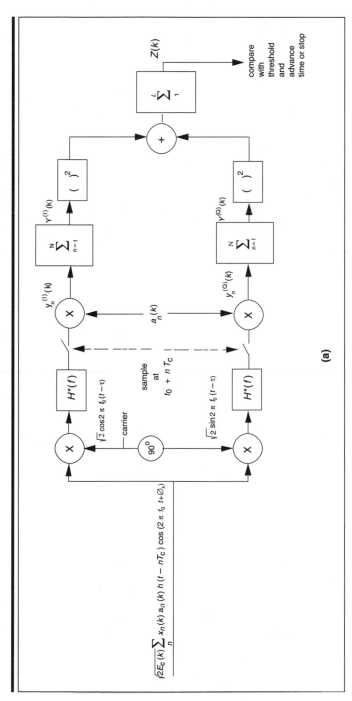

Figure 3.1 (a) Time hypothesis testing for BPSK spreading.

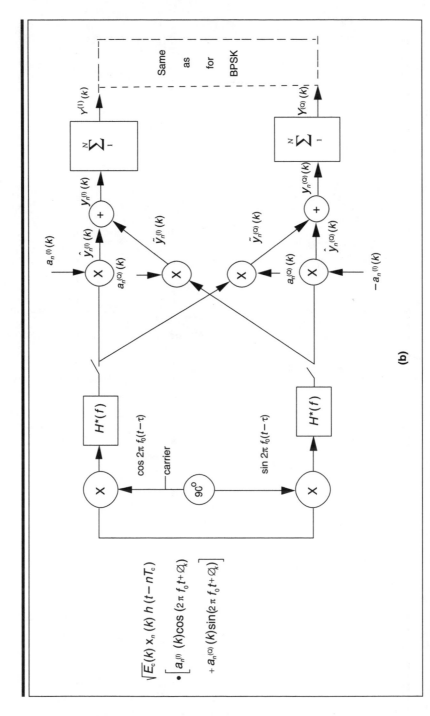

Figure 3.1   (b) Time hypothesis testing for QPSK spreading.

the first-order statistics of the hypothesis testing device output components for both BPSK and QPSK modulation in additive Gaussian interference.

We may take the signal to be unmodulated by data, as would be the case for a pilot or synchronization sequence. However, if the signal is data-modulated, but the data sequence is synchronized to the pseudorandom sequence, the data rate is $(1/N)$th of the chip rate ($N = 1/RT_c$ so that the data sequence remains constant over N chips). Then, without loss of generality, because of the squaring operation, we may let

$$x_n(k) = +1 \qquad \text{for all } n = 1, 2, \ldots, N. \qquad (3.1)$$

Following the same steps as in Sections 2.3 and 2.4, it is easily established that with the timing error, τ, the means of $Y^{(I)}(k)$ and $Y^{(Q)}(k)$ are

$$E[Y^{(I)}(k)] = N\sqrt{E_c(k)} \, (\cos \phi_k) R(\tau),$$

$$(3.2)$$

$$E[Y^{(Q)}(k)] = N\sqrt{E_c(k)} \, (\sin \phi_k) R(\tau),$$

where

$$R(\tau) = \int_{-\infty}^{\infty} |H(f)|^2 \cos(2\pi f\tau) \, df.$$

As shown in Section 2.3.3 for the cases of strictly time-limited and band-limited filters,

$$R(\tau) = \begin{cases} 1 - |\tau|/T_c, & |\tau| \le T_c, \\ 0, & \text{otherwise,} \end{cases} \qquad \text{(time-limited)}, \qquad (3.3a)$$

$$R(\tau) = \frac{\sin(\pi\tau/T_c)}{(\pi\tau/T_c)} \qquad \text{(bandlimited)}. \qquad (3.3b)$$

Note also that the phase ϕ_k is immaterial because the terms of (3.2) are squared and added by the device of Figure 3.1a.

The variance, also derived in Section 2.3, is the sum of the contribution of other-user interference, background noise and interchip interference. We ignore the last as negligible compared to the other two. It then follows from (2.31) and (2.18) that

$$\text{Var}(Y^{(I)}) = \text{Var}(Y^{(Q)}) = NI_0/2, \qquad (3.4)$$

where

$$I_0 = N_0 + \sum_{j \neq k} E_c(j) \int |H(f)|^4 \, df/T_c.$$ (3.5)

The integral in (3.5) equals $\frac{2}{3}$ for a time-limited waveform and equals 1 for an ideal bandlimited waveform, according to (2.26) and (2.27), respectively.

Actually the means and variances, as given by (3.2) through (3.5), are based on two implicit assumptions:

(a) The phase ϕ_k remains constant over the N chips accumulated, which implies that frequency error is negligible;

(b) N corresponds to one or more periods of the pseudorandom sequences.

We shall investigate both these issues and determine how the means and variances are affected when the assumptions do not hold. First, however, we consider the basic hypothesis testing device for QPSK spreading.

3.2.2 Hypothesis Testing for QPSK Spreading

To accommodate the effect of QPSK spreading, the detection device of Figure 3.1a must be modified to include the cross-arms shown in Figure 3.1b. We denote the direct arm terms by $\hat{}$ and the cross-arm terms by $\tilde{}$, as shown on the figure, and drop the user-index k for notational simplicity. We then have for each chip for both sets of terms,

$$\hat{y}_n^{(I)} = \sqrt{E_c}\, [R(\tau)/2](\cos \phi + a_n^{(Q)} a_n^{(I)} \sin \phi) + \hat{v}_n^{(I)},$$

$$\tilde{y}_n^{(I)} = \sqrt{E_c}\, [R(\tau)/2](\cos \phi - a_n^{(I)} a_n^{(Q)} \sin \phi) + \tilde{v}_n^{(I)},$$

$$\hat{y}_n^{(Q)} = \sqrt{E_c}\, [R(\tau)/2](\sin \phi - a_n^{(Q)} a_n^{(I)} \cos \phi) + \hat{v}_n^{(Q)},$$

$$\tilde{y}_n^{(Q)} = \sqrt{E_c}\, [R(\tau)/2](\sin \phi + a_n^{(I)} a_n^{(Q)} \cos \phi) + \tilde{v}_n^{(Q)},$$

where the v terms represent the effect of noise and interference. The I and Q pairs are summed to obtain

$$y_n^{(I)} = \hat{y}_n^{(I)} + \tilde{y}_n^{(I)} = \sqrt{E_c}\, R(\tau) \cos \phi + v_n^{(I)},$$

$$y_n^{(Q)} = \hat{y}_n^{(Q)} + \tilde{y}_n^{(Q)} = \sqrt{E_c}\, R(\tau) \sin \phi + v_n^{(Q)}.$$

After summation over N chips, with the same two assumptions [(a) and (b)] made for BPSK spreading, we have

$$E[Y^{(I)}] = N\sqrt{E_c}\, R(\tau)\cos\phi,$$

$$E[Y^{(Q)}] = N\sqrt{E_c}\, R(\tau)\sin\phi,$$

where

$$R(\tau) = \int |H(f)|^2 \cos(2\pi f\tau)\, df.$$

The variances are also the same as for BPSK, since the cross-term summation replaces the multiplication by $\sqrt{2}$ in Figure 3.1a (doubling power). Thus,

$$\mathrm{Var}(Y^{(I)}) = \mathrm{Var}(Y^{(Q)}) = NI_0/2,$$

where

$$I_0 = N_0 + \sum_{j\neq k} E_c(j) \int |H(f)|^4\, df/T_c.$$

All of these are the same as Equations (3.2), (3.4), and (3.5) for BPSK.

3.2.3 Effect of Frequency Error

Suppose that the phase is not constant because there is an error Δf in carrier frequency between transmitter and receiver. Then assuming no timing error, $\tau = 0$, Equations (3.2) are replaced by

$$E[Y^{(I)}] = \sum_{n=1}^{N} \sqrt{E_c}\, \Phi\{\cos[2\pi(\Delta f)t + \phi]\}\,|_{t=nT_c},$$

$$E[Y^{(Q)}] = \sum_{n=1}^{N} \sqrt{E_c}\, \Phi\{\sin[2\pi(\Delta f)t + \phi]\}\,|_{t-nT_c},$$

where $\Phi\{\ \}$ is the operation of filtering by the matched filter $H^*(f)$. It is clear that, since the variances are independent of the kth user's signal, they are unaffected by the frequency error.

For simplicity, we examine the two cases of time-limited and band-limited filters and show that for large N, the two degradations are nearly the same. We surmise from this that the result will be nearly the same for all practical filters. For the time-limited waveform, the filter performs an integration over the chip. Thus,

$$
\begin{aligned}
E[Y^{(I)}] &= \sqrt{E_c} \sum_{n=1}^{N} \int_{(n-1)T_c}^{nT_c} \cos[2\pi(\Delta f)t + \phi]\, dt/T_c \\
&= \sqrt{E_c} \int_{0}^{NT_c} \cos[2\pi(\Delta f)t + \phi]\, dt/T_c \\
&= \sqrt{E_c} \left[\frac{\sin[2\pi N(\Delta f)T_c]}{2\pi\, \Delta f T_c} \cos\phi - \frac{1 - \cos[2\pi N(\Delta f)T_c]}{2\pi\, \Delta f T_c} \sin\phi \right].
\end{aligned}
$$

Similarly,

$$
\begin{aligned}
E[Y^{(Q)}] &= \sqrt{E_c} \int_{0}^{NT_c} \sin[2\pi(\Delta f)t + \phi]\, dt/T_c \\
&= \sqrt{E_c} \left[\frac{1 - \cos(2\pi N(\Delta f)T_c)}{2\pi\, \Delta f T_c} \cos\phi + \frac{\sin(2\pi N\Delta f T_c)}{2\pi\, \Delta f T_c} \sin\phi \right].
\end{aligned}
$$

Since, as will be discussed in the next section, performance depends only on the sum

$$
Z = [Y^{(I)}]^2 + [Y^{(Q)}]^2,
$$

it follows that the noiseless signal-only terms of Z are

$$
\begin{aligned}
\{E[Y^{(I)}]\}^2 + \{E[Y^{(Q)}]\}^2 &= N^2 E_c \left\{ \left[\frac{\sin(2\pi N\Delta f T_c)}{2\pi N\Delta f T_c} \right]^2 \right. \\
&\qquad \left. + \left[\frac{1 - \cos(2\pi N\Delta f T_c)}{2\pi N\Delta f T_c} \right]^2 \right\} \\
&= N^2 E_c \left[\frac{2 - 2\cos(2\pi N\Delta f T_c)}{(2\pi N\Delta f T_c)^2} \right] \\
&= N^2 E_c \left[\frac{\sin(\pi N\Delta f T_c)}{\pi N\Delta f T_c} \right]^2.
\end{aligned}
$$

Thus, the degradation to the signal due to a frequency error of Δf hertz is

$$
D(\Delta f) = \left[\frac{\sin(\pi N\Delta f T_c)}{\pi N\Delta f T_c} \right]^2 \qquad \text{(time-limited).} \qquad (3.6)
$$

Consider now the ideal bandlimited case (more precisely, a physically realizable approximation thereof). If $|\Delta f| \ll W/2$, or equivalently $|\Delta f T_c| \ll 1/2$, the filter output may be approximated by a delayed version of the input. Thus,

$$
\begin{aligned}
E[Y^{(I)}] &\approx \sqrt{E_c} \sum_{n=1}^{N} \cos(2\pi n\Delta f T_c + \phi) \\
&= \sqrt{E_c} \sum_{n=1}^{N} \cos(2\pi n\Delta f T_c) \cos\phi - \sum_{n=1}^{N} \sin(2\pi n\Delta f T_c) \sin\phi
\end{aligned}
$$

and

$$
E[Y^{(Q)}] \approx \sqrt{E_c} \sum_{n=1}^{N} \sin(2\pi n\Delta f T_c) \cos\phi + \sum_{n=1}^{N} \cos(2\pi n\Delta f T_c) \sin\phi,
$$

whence

$$
\begin{aligned}
\{E[Y^{(I)}]\}^2 + \{E[Y^{(Q)}]\}^2 &\approx E_c \left\{ \left[\sum_{n=1}^{N} \sin(2\pi n\Delta f T_c) \right]^2 \right. \\
&\quad \left. + \left[\sum_{n=1}^{N} \cos(2\pi n\Delta f T_c) \right]^2 \right\} \\
&= E_c \left[\frac{\sin(\pi N\Delta f T_c)}{\sin(\pi \Delta f T_c)} \right]^2, \qquad |\Delta f T_c| < \tfrac{1}{2}.
\end{aligned}
$$

But for $N \gg 1$ and $\pi(\Delta f)T_c \ll 1$, the degradation due to frequency error is

$$
\begin{aligned}
D(\Delta f) &\approx \frac{1}{N^2} \left[\frac{\sin(\pi N\Delta f T_c)}{\sin(\pi \Delta f T_c)} \right]^2 \\
&\approx \left[\frac{\sin(\pi N\Delta f T_c)}{\pi N\Delta f T_c} \right]^2 \qquad \text{(bandlimited)}.
\end{aligned}
\tag{3.7}
$$

This approximation is the same as the exact expression for the time-limited case.

In either case, as long as $N\Delta f T_c$ is relatively small, the degradation will be small. For example, suppose the chip time $T_c = 1$ microsecond (spreading bandwidth ≈ 1 MHz) and the frequency error $\Delta f = 1$ kHz. Then for $N = 100$, the degradation is only .97 (-0.14 dB). For $N = 250$, it becomes .81 (-0.91 dB).

3.2.4 Additional Degradation When *N* Is Much Less Than One Period

We now examine the effect of dropping assumption (b) of Section 3.2.1. When the received and transmitted sequences are perfectly synchronized, the desired user sequence contributes only to the mean. When they are not synchronized—meaning that the receiver clock (index n) is off by more than one chip, but N is one or more periods of the pseudorandom sequence—the contribution of the desired user is nearly zero, according to the delay-and-add property [(2.10) of Section 2.2]. However, it is clear from the preceding that N generally must be much less than one period. This is necessary both to maintain a tolerable degradation due to frequency error and to satisfy condition (3.1) that the data $x_n(k)$ are constant over the N chips of the summation.

In the event that N is much smaller than the pseudorandom sequence period, when $a_n(k)$ is out of synchronization it will appear like a random "coin-flipping" sequence. Its mean will be zero, but the sum will contribute to the variance an amount proportional to $NE_c(k)$. Thus, the variance equations (3.4) and (3.5) must be increased by the *term previously excluded in the summation* of (3.5).

That is,

$$\text{Var}(Y^{(I)}) = \text{Var}(Y^{(Q)}) = NI_0/2,$$

where now

$$I_0 = N_0 + \sum_{j=1}^{k_u} E_c(j) \int |H(f)|^4 \, df/T_c. \tag{3.8}$$

While (3.8) applies only to the unsynchronized case, for simplicity we shall apply it everywhere (as an upper bound in the synchronized case). When the number of users k_u is large, the additional term has minimal effect.

3.3 Detection and False Alarm Probabilities

Up until now we have considered only the first- and second-order statistics of the hypothesis testing devices. Now we show how these may be used to distinguish between two cases: when the timing error is less than one chip time, so that the Y variables have a positive mean; and when the error is greater than one chip time, so that the Y variables have negligible or zero means. In Appendix 3A, we also justify the processing functions and indicate under what conditions they are optimal or moderately sub-optimal.

3.3.1 Fixed Signals in Gaussian Noise ($L = 1$)

We now determine the complete first-order statistics of the detection variable $Z = [Y^{(I)}]^2 + [Y^{(Q)}]^2$ as shown in Figure 3.1. With $L = 1$ and precisely known center frequency, f_0, and chip timing, it is shown in Appendix 3A that the hypothesis testing device is optimum for a signal with unknown phase in Gaussian noise, according to either the Bayes or the Neyman–Pearson optimality criteria [Neyman and Pearson, 1933; Helstrom, 1968]. In the latter case, the threshold θ is chosen to establish a specified false alarm probability, P_F. This then guarantees that the detection probability, P_D, is maximized for that P_F level. It is also shown in Appendix 3A.3 that for $L = 1$ in Figure 3.1, the false alarm and detection probabilities are obtained as integrals of two likelihood functions: $p_0(Z)$, under the assumption that the hypothesis is incorrect; and $p_1(Z)$, under the assumption that it is correct. For both BPSK and QPSK spreading, these are given by

$$p_0(Z) = \frac{e^{-Z/V}}{V},$$

$$p_1(Z) = \frac{\exp[-(Z + M_D^2)/V]}{V} \, \mathcal{I}_0\left(\frac{2\sqrt{M_D^2 Z}}{V}\right).$$

(3.9)

Here, V is twice the variance of each component, and $\mathcal{I}_0(\ )$ is the zeroth-order modified Bessel function—not to be confused with the interference density[1] value I_0. M_D^2 is the mean square, which is obtained as the sum of the squares of the I and Q component means of (3.2) (see Figure 3.1), which includes the effect of the timing error.

Thus, from (3.2) and (3.8) we have

$$M_D^2 = N^2 E_c(k)R^2(\tau), \qquad \tau < 1, \tag{3.10a}$$

$$V = NI_0 = N\left(N_0 + \sum_{j=1}^{k_u} E_c(j)\int |H(f)|^4\, df/T_c\right). \tag{3.10b}$$

---

[1] The two will be distinguished throughout by the fact that the latter is a parameter value, while the former is a function $\mathcal{I}_0(\ )$ and is written using a script font.

From the likelihood functions of (3.9), the false alarm and detection probabilities are obtained as

$$P_F = \int_\theta^\infty p_0(Z)\, dZ = e^{-\theta/V}, \tag{3.11}$$

$$P_D = \int_\theta^\infty p_1(Z)\, dZ = \int_{\theta/V}^\infty e^{-(x+\mu)} \mathcal{J}_0(2\sqrt{\mu x})\, dx. \tag{3.12}$$

The latter integral is the Marcum Q-function, and we have defined the SNR

$$\mu = \frac{M_D^2}{V} = \frac{N\, E_c(k) R^2(\tau)}{N_0 + \sum\limits_{j=1}^{k_u} E_c(j) \int |H(f)|^4\, df/T_c}. \tag{3.10c}$$

Figure 3.2a shows the detection probability as a function of false-alarm probability for a number of values of the SNR, μ. In the detection theory literature [Helstrom, 1968], these are known as receiver operating characteristics.

3.3.2 Fixed Signals in Gaussian Noise with Postdetection Integration ($L > 1$)

As noted in Section 3.2, the value of N is limited by two requirements: that the frequency error, Δf, be sufficiently small so that $N(\Delta f)T_c < 1$, and that the data modulation $x_n(k)$ remain at the same sign throughout the N chips. Thus, to dwell longer than N chips at a given hypothesis, the energy must be accumulated over successive N-chip intervals and then added as shown in Figures 3.1a and 3.1b. In this case, for $L > 1$, the false alarm and detection are generalized to (see Appendix 3A.4)

$$P_F = \int_\theta^\infty p_0(Z)\, dZ = \int_{\theta/V}^\infty \frac{x^{L-1} e^{-x}}{(L-1)!}\, dx = e^{-\theta/V} \sum_{k=0}^{L-1} \frac{(\theta/V)^k}{k!}, \tag{3.13}$$

$$P_D = \int_\theta^\infty p_1(Z)\, dZ = \int_{\theta/V}^\infty \left(\frac{x}{L\mu}\right)^{(L-1)/2} e^{-(x+L\mu)} \mathcal{J}_{L-1}(2\sqrt{L\mu x})\, dx. \tag{3.14}$$

Here, $\mathcal{J}_{L-1}(\ )$ is the $(L-1)$th-order modified Bessel function and μ is the SNR per N-chip interval (3.10c). The latter integral is the Lth-order Marcum Q-function [Marcum, 1950; Helstrom, 1968]. The case for $L = 2$ is shown in Figure 3.2a.

3.3.3 Rayleigh Fading Signals ($L \geq 1$)

Often in terrestrial propagation, with multiple reflections and refractions arriving at the receiver with minimal relative delay and random phases (as will be described in Section 4.3), the signal is Rayleigh fading. In such a case, its I and Q components are Gaussian, as well as the noise. We assume

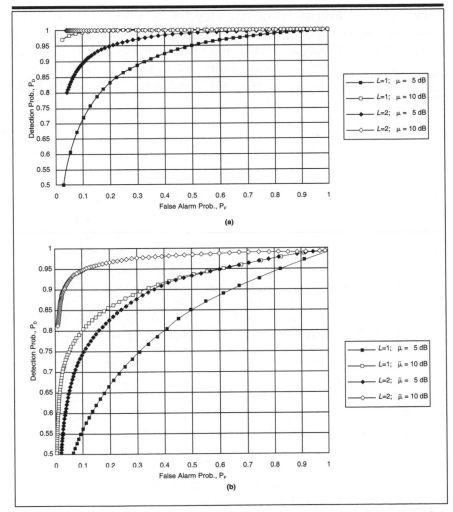

Figure 3.2 (a) Detection and false alarm probabilities for unfaded channel, $L = 1$ and 2. (b) Detection and false alarm probabilities for Rayleigh fading, $L = 1$ and 2.

that the Rayleigh fading is slow enough that the amplitude as well as the phase (and hence I and Q components) remain constant over N chip times but fast enough that successive N chip segments are essentially independent.[2] While the false alarm probability does not change, the likelihood function for the correct hypothesis is shown in Appendix 3A.5 to be the same as $p_0(Z)$, but with V replaced by

$$V_F = N[I_0 + N\overline{E_c(k)}R^2(\tau)] = V(1 + \overline{\mu}), \qquad (3.15)$$

where $\overline{E_c}$ is the average energy per chip for the Rayleigh faded signal, since now the signal as well as the noise components are Gaussian. V and μ are defined by (3.10b) and (3.10c), respectively. In this case the detection probability expression becomes the same as the false alarm probabilities in (3.11) and (3.13), but with V replaced by the larger V_F (Appendix 3A.5). Thus,

$$P_D = e^{-\theta/V_F} \sum_{k=0}^{L-1} \frac{(\theta/V_F)^k}{k!}. \qquad (3.16)$$

The receiver operating characteristics for Rayleigh fading signals are shown in Figure 3.2b for $L = 1$ and 2. Note that the performance for $\overline{\mu} = 5$ dB with $L = 2$ (for a total signal-to-noise ratio of 8 dB) is about the same as for $\overline{\mu} = 10$ dB with $L = 1$. This implies that the twofold diversity gains about 2 dB for this fading example. We shall encounter further cases of diversity improvements in fading in subsequent chapters.

3.4 The Search Procedure and Acquisition Time

Until now we have only considered static tests for the pseudorandom sequence timing. Clearly, with no prior information, all P initial vector hypotheses must be tested, where P is the period of the MLSR-generated sequence. Even this may not be sufficient: If the sequence is tested once for every chip time, the correct hypothesis may be tested with a one-half chip timing error. In that case $R^2(\tau)$ may be low by a significant factor and hence cause a several-decibel loss in performance [6 dB for time-limited waveforms (3.3a), 4 dB for bandlimited waveforms (3.3b)]. Thus, each chip hypothesis should be sampled several times per chip. With a bandlimited, or nearly so, waveform, with ℓ hypotheses per chip the band-

---

[2] If the latter assumption is not valid, the advantage of post-detection integration is reduced.

limited waveform loss is reduced to $\sin^2[\pi/(2\ell)]/[\pi/(2\ell)]^2$ (0.9 dB loss for $\ell = 2$, 0.2 db for $\ell = 4$). To test all hypotheses over the whole period will then require a search over $\nu = \ell P$ possible hypotheses. Note that to ensure sufficiently high detection probabilities for acceptable false alarm probabilities as determined in the last section, each test requires summation over NL chips. Thus, the total search time will be proportional to νNL chip times. Of course, if the search is divided equally among k parallel processors, then the number of states that may need to be tested by each processor is ν/k.

In most practical systems, the entire pseudorandom sequence does not need to be tested. The receiver will have a reasonable estimate of the correct time, and the initial times of all user sequences are known. The call initiator may also inform the receiver of its intention to transmit, sending along its identifying initial vector and generator polynomial using an auxiliary channel, possibly spread by a shorter sequence. In such cases, the parameter P will refer to the number of chip hypotheses to be tested rather than the period of the pseudorandom sequence.

3.4.1 Single-Pass Serial Search (Simplified)

The acquisition parameter of greatest interest is, of course, the total search time required to find the correct timing hypothesis, meaning to within less than T_c. Note that if the search has proceeded through all possibilities without accepting the correct hypothesis, the process will repeat. Hence, a serial search can be described by a circular diagram, as shown in Figure 3.3. Here each node represents a hypothesis. The labels on branches between nodes indicate the probability of the particular transition, multiplied by a power of the variable z. The power is used to indicate the hypothesis testing time periods required for making the given transition. (Each hypothesis testing period corresponds to NL chip times).

We consider first an idealized "genie-aided" situation, where the chip sampling occurs at the peak $[R(\tau) = 1]$, and thus only one sample per chip suffices ($\nu = P$). The node at the top of the state diagram of Figure 3.3a represents the correct state with perfect timing. All other states on the outer circle represent the $P - 1$ incorrect hypotheses. States on the inner dotted circle represent false alarm states reached as a consequence of the acceptance of an incorrect hypothesis. We assume that such an incorrectly chosen time hypothesis will eventually be excluded, but only after K testing periods, during which an auxiliary device recognizes that the system timing is still not locked. Note that the state where the search is begun, can be any one of the ν nodes on the outer circle, although for later convenience we exclude starting in the correct timing state.

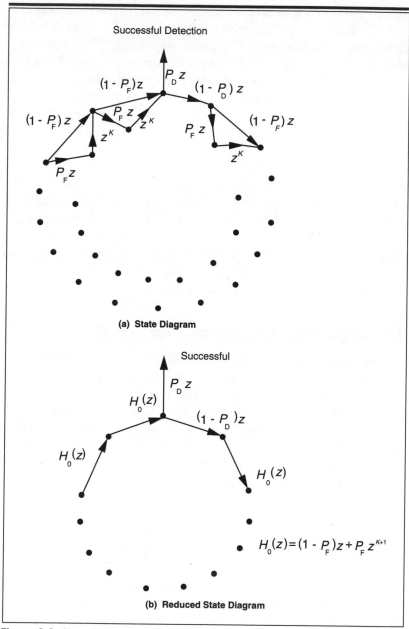

Figure 3.3 Simplified search state diagrams. (a) State diagram. (b) Reduced state diagram.

The search time is a random variable equal to the sum of the transition times of all the branches on the path taken in the state diagram, from any one of the equally likely initial states to the final correct detection state at the top of the diagram. It is relatively simple to obtain a generating function for the distribution as the transfer function from each potential starting node (on the outer circle) to the final destination node. This is simplified further by noting that any excursion into the inner circle of nodes ultimately returns to the next outer circle node with a delay of $K + 1$. Thus, for the purpose of computing the transfer function, only the nodes of the outer circle need be considered, connected by branch transfer functions

$$H_0(z) = (1 - P_F)z + P_F z^{K+1}. \tag{3.17}$$

This applies to all cases except the branches coming from the node at the top of the circle (which connects also to the final node), which remain unchanged (Figure 3.3b). Then the transfer function from an initial node that is i branches counterclockwise from the top to the final destination (correct) node is

$$U_i(z) = \frac{H_0^i(z)P_D z}{1 - (1 - P_D)zH_0^{\nu-1}(z)}.$$

Now, since all nodes are a priori equally likely, the total transfer function averaged over all $\nu - 1$ starting nodes (assuming we start anywhere but the correct state) is

$$U(z) = \frac{1}{\nu - 1} \sum_{i=1}^{\nu-1} U_i(z) = \frac{1}{\nu - 1} \frac{P_D z \sum_{i=1}^{\nu-1} H_0^i(z)}{1 - (1 - P_D)zH_0^{\nu-1}(z)}$$
$$= \frac{P_D z H_0(z)[1 - H_0^{\nu-1}(z)]}{(\nu - 1)[1 - H_0(z)][1 - (1 - P_D)zH_0^{\nu-1}(z)]}. \tag{3.18}$$

$H_0(z)$ is given by (3.17). The generating function is obtained by expanding (3.18) through polynomial division. Denote this

$$U(z) = \sum_{k=1}^{\infty} U_k z^k.$$

Then the *average time to acquisition*, denoted $\overline{T}_{ACQ}$, is given by[3]

$$\overline{T}_{ACQ} = \sum_{k=1}^{\infty} k\,U_k = \frac{dU(z)}{dz}\bigg|_{z=1} \tag{3.19}$$

---

[3] L'Hospital's rule is needed to evaluate the limit as z approaches 1, since both numerator and denominator approach 0.

test periods. Note also that taking the second derivative yields the difference between the second and the first moment, since

$$\left.\frac{d^2U(z)}{dz^2}\right|_{z=1} = \sum_{k=1}^{\infty} k(k-1)U_k = \overline{T^2_{ACQ}} - \overline{T}_{ACQ}.$$

Hence, it follows that the variance of the time to acquisition is

$$\text{Var}(T_{ACQ}) = \overline{T^2_{ACQ}} - (\overline{T}_{ACQ})^2$$
$$= \left.\left\{\frac{d^2U(z)}{dz^2} + \frac{dU(z)}{dz}\left[1 - \frac{dU(z)}{dz}\right]\right\}\right|_{z=1}. \qquad (3.20)$$

Note that each test takes NLT_c seconds. $\overline{T}_{ACQ}$ must be multiplied by this amount to obtain the mean time in seconds, and the variance must be multiplied by its square. The detection and false alarm probabilities required in (3.17) and (3.18), as well as in what follows, are computed from the expressions given in Section 3.3 (or are obtained from Figures 3.2a or b).

3.4.2 Single-Pass Serial Search (Complete)

Consider now the actual situation in which over each chip time T_c, multiple timing hypotheses are serially tested for timing separations T_c/ℓ, and the nodes are increased ℓ-fold so now $v = \ell P$. All $v - 2\ell$ states that can lead to false alarms are treated exactly as before. However, the 2ℓ states within one chip of the correct timing appear in the state diagram as shown in Figure 3.4a, where P_{Dj} is the detection probability expression [e.g., (3.12) with parameter M_D^2 of (3.10a) evaluated at $\tau = \tau_j$], and where

$$|\tau_{j+1} - \tau_j| = T_c/\ell, \qquad j \text{ any integer.}$$

The worst case corresponds to sampling times that differ from the correct (peak) time by

$$\frac{j - (\ell + 1/2)}{\ell} T_c, \qquad j = 1, 2, \ldots, 2\ell.$$

Combining all 2ℓ states into one leads to the simplified diagram of Figure 3.4b with transfer functions of branches emanating from this node,

$$H_D(z) = \sum_{j=1}^{2\ell} P_{Dj} z \prod_{i=1}^{j-1} [(1 - P_{D_i})z], \qquad (3.21)$$

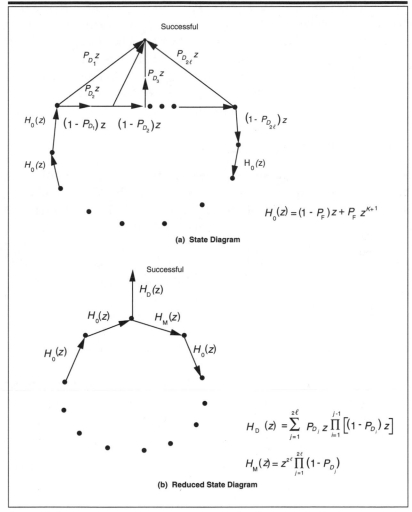

Figure 3.4 Complete state diagram. (a) State diagram. (b) Reduced state diagram.

and

$$H_{\mathrm{M}}(z) = z^{2\ell} \prod_{j=1}^{2\ell} (1 - P_{D_j}). \qquad (3.22)$$

Thus, assuming we start anywhere outside the correct chip period, the transfer function for this general serial search is

$$U(z) = \frac{H_0(z)H_{\mathrm{D}}(z)[1 - H_0^{\nu-2\ell}(z)]}{(\nu - 2\ell)[1 - H_0(z)][1 - H_{\mathrm{M}}(z)H_0^{\nu-2\ell}(z)]}. \qquad (3.23)$$

This specializes to (3.18) for $H_D(z) = P_D z$ and $H_M(z) = (1 - P_D)z$, with 2ℓ replaced by 1. (Note that the previous case applied to sampling exactly at the peak; when there is a sampling error, there will be at least two samples within one chip.) $\overline{T}_{ACQ}$ and $\mathrm{Var}(T_{ACQ})$ are obtained as before from (3.19) and (3.20). It is clear that the mean-time-to-acquisition parameter can be minimized by appropriate choice of θ (upon which P_D and P_F depend) and of parameters N and L, since the mean time in seconds is $(NLT_c)\,\overline{T}_{ACQ}$.

3.4.3 Multiple Dwell Serial Search

Besides the choice of parameters, another approach to minimizing the serial search is to perform successive tests with multiple dwell times. The flow chart of Figure 3.5 illustrates this concept. Each hypothesis is first checked for a time $L_1 N T_c$ against a θ_1 threshold. If it fails, the system proceeds to check the next hypothesis. If it succeeds, a larger dwell $L_2 N T_c$ is checked against threshold θ_2, where $m = L_2/L_1 > 1$ and $\theta_2 > \theta_1$. If this succeeds, the hypothesis is deemed correct (although it may be a false alarm as determined over K later time periods). If it fails the second threshold, it is discarded and the next hypothesis is checked. Quickly eliminating unlikely hypotheses significantly reduces the overall acquisition time, with only an occasional increase because a correct hypothesis is quickly eliminated in the first pass.

The state diagram is the same as for the single dwell time Figure 3.4b, but with branch transfer functions defined as follows, with superscripts (1) and (2) indicating the first and second dwell.

For incorrect hypothesis branches,

$$\tilde{H}_0(z) = z(1 - P_F^{(1)}) + z^{1+m}P_F^{(1)}(1 - P_F^{(2)}) + z^{1+m+K}P_F^{(1)}P_F^{(2)}. \quad (3.24)$$

For the correct-hypothesis-missed branch,

$$\tilde{H}_M(z) \leq H_M^{(1)}(z) + H_D^{(1)}(z)H_M^{(2)}(z^m), \quad (3.25)$$

and for the correct-hypothesis-detected branch,

$$\tilde{H}_D(z) \leq H_D^{(1)}(z)H_D^{(2)}(z^m). \quad (3.26)$$

Here the upper bound represents a possible over-estimate of the time, based on reexamining all 2ℓ correct hypotheses on the second pass, although the first threshold is passed on only one of these.

Otherwise, the procedure for computing acquisition time statistics is the same as for the single dwell case. That is, for the two-dwell case, upper

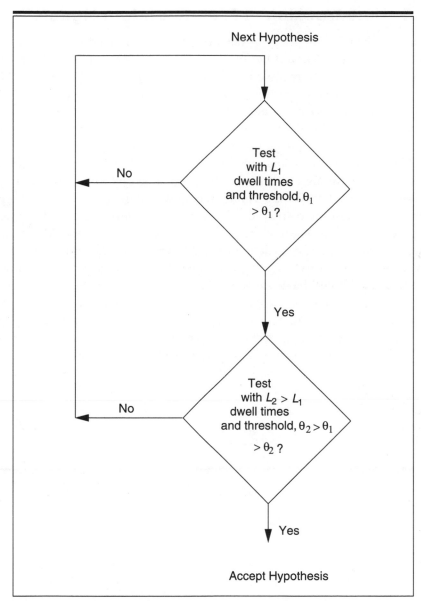

Figure 3.5 Two-dwell serial search.

bounds on the mean and variance of the acquisition time are obtained by replacing the $H(z)$ terms by $\tilde{H}(z)$ terms in (3.23). The ultimate advantage of the two-dwell search (which may be generalized to multiple dwells) depends on the application and on the optimization of parameters [Polydoros and Simon, 1984; Simon *et al.*, 1985].

3.5 Time Tracking of Pseudorandom Signals

Once receiver timing has been synchronized to within a fraction of a chip time, the estimate should be further refined to approach zero. Furthermore, because of the relative motion of transmitter and receiver and the instability of clocks, corrections must be made continuously. This process, which is called *tracking,* is performed by the so-called "early–late" gate device [Spilker, 1963] of Figure 3.6. It is an elaboration of the basic demodulator and acquisition device of Figures 3.1a and b. Recall from (3.10a) that the signal component of the energy Z is proportional to $R^2(\tau)$ where τ is the timing error. We perform the same operation advanced by Δ seconds and delayed Δ seconds, where $\Delta < T_c$ is a fraction of a chip time. Then the time-displaced energy measurements, Z_- and Z_+, have signal components M_D^2 that are proportional to $R^2(\tau - \Delta)$ and $R^2(\tau + \Delta)$, respectively. Since $R(\tau)$, as defined in (3.2), is an even function, monotonically decreasing in τ within the main lobe, the difference of the measurements $Z_- - Z_+$ will have a mean proportional to $R^2(\tau - \Delta) - R^2(\tau + \Delta)$. This will be positive for $\tau > 0$ (late estimate) and negative for $\tau < 0$ (early estimate). The measurement $Z_- - Z_+$ will be used to correct timing by means of a tracking loop technique. We will defer consideration of the loop performance to the next subsection. We first determine the statistics of the early–late gate measurements. Throughout we shall use the notation

$$R_+ = R(\tau + \Delta),$$
$$R_- = R(\tau - \Delta).$$

$$(3.27)$$

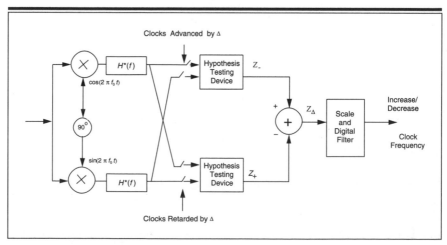

Figure 3.6 Early–late gate.

3.5.1 Early-Late Gate Measurement Statistics

We consider first Z_+. The statistics of Z_- will be the same, but with R_+ replaced by R_-. We assume as before that the number of users k_u and the number of accumulated chips N are sufficiently large that we may take $Y_+^{(I)}$ and $Y_+^{(Q)}$ to be Gaussian random variables according to the central limit theorem approximation. Then we may express each as

$$Y_+^{(I)} = \sum_{n=1}^{N} \left[\sqrt{E_c}\, R_+ \cos \phi + \nu_n^{(I)} \right],$$
$$Y_+^{(Q)} = \sum_{n=1}^{N} \left[\sqrt{E_c}\, R_+ \sin \phi + \nu_n^{(Q)} \right]. \tag{3.28}$$

$\nu_n^{(I)}$ and $\nu_n^{(Q)}$ are independent Gaussian random variables with zero means and

$$\mathrm{Var}[\nu_n^{(I)}] = \mathrm{Var}[\nu_n^{(Q)}] = I_0/2. \tag{3.29}$$

We then obtain the mean and variance of the late measurement Z_+, by a direct but tedious computation, as follows:

$$
\begin{aligned}
E(Z_+) &= E\{[Y_+^{(I)}]^2\} + E\{[Y_+^{(Q)}]^2\} \\
&= N^2 E_c R_+^2 \cos^2 \phi + E\left\{ \left[\sum_{n=1}^{N} \nu_n^{(I)} \right]^2 \right\} \\
&\quad + N^2 E_c R_+^2 \sin^2 \phi + E\left\{ \left[\sum_{n=1}^{N} \nu_n^{(Q)} \right]^2 \right\} \\
&= N^2 E_c R_+^2 + 2N(I_0/2) = N(NE_c R_+^2 + I_0),
\end{aligned}
\tag{3.30}
$$

$$
\begin{aligned}
E(Z_+^2) &= E(\{[Y_+^{(I)}]^2 + [Y_+^{(Q)}]^2\}^2) \\
&= E\{[Y_+^{(I)}]^4\} + E\{[Y_+^{(Q)}]^4\} + 2E\{[Y_+^{(I)}]^2\}E\{[Y_+^{(Q)}]^2\}.
\end{aligned}
\tag{3.31}
$$

This follows from the fact that all $\nu^{(I)}$ and $\nu^{(Q)}$ variables are mutually independent. The squared terms in (3.31) are as above:

$$E\{[Y_+^{(I)}]^2\} = N^2 E_c R_+^2 \cos^2 \phi + NI_0/2,$$
$$E\{[Y_+^{(Q)}]^2\} = N^2 E_c R_+^2 \sin^2 \phi + NI_0/2. \tag{3.32}$$

For the fourth-power terms, since odd moments of ν are zero,

$$E\{[Y_+^{(I)}]^4\} = N^4 E_c^2 R_+^4 \cos^4 \phi$$
$$+ 6N^3(I_0/2)E_c R_+^2 \cos^2 \phi + E\left\{\left[\sum_{n=1}^{N} \nu_n^{(I)}\right]^4\right\},$$

$$E\{[Y_+^{(Q)}]^4\} = N^4 E_c^2 R_+^4 \sin^4 \phi \tag{3.33}$$
$$+ 6N^3(I_0/2)E_c R_+^2 \sin^2 \phi + E\left\{\left[\sum_{n=1}^{N} \nu_n^{(Q)}\right]^4\right\}.$$

But since all the ν variables are independent and Gaussian with zero means,

$$E\left\{\left[\sum_{n=1}^{N} \nu_n^{(I)}\right]^4\right\} = 3\left\{E\sum_{n=1}^{N} [\nu_n^{(I)}]^2\right\}^2 \tag{3.34}$$
$$= 3(NI_0/2)^2.$$

Hence, combining (3.31) through (3.34), we have

$$E(Z_+^2) = N^4 E_c^2 R_+^4 (\cos^4 \phi + \sin^4 \phi) + 3N^3 I_0 E_c R_+^2 + 3(NI_0)^2/2$$
$$+ 2N^4 E_c^2 R_+^4 \cos^2 \phi \sin^2 \phi + (NI_0)^2/2 + N^3 I_0 E_c R_+^2 \tag{3.35}$$
$$= N^4 E_c^2 R_+^4 + 2[N^2 I_0^2 + 2N^3 I_0 E_c R_+^2].$$

Finally, subtracting the square of (3.30) from (3.35), we obtain

$$\begin{aligned} \text{Var}(Z_+) &= E(Z_+^2) - [E(Z_+)]^2 \\ &= N^4 E_c^2 R_+^4 + 2[N^2 I_0^2 + N^3 I_0 E_c R_+^2] \\ &\quad - [N^4 E_c^2 R_+^4 + N^2 I_0^2 + 2N^3 I_0 E_c R_+^2] \\ &= (NI_0)^2 [1 + 2N(E_c/I_0)R_+^2]. \end{aligned} \tag{3.36}$$

The same calculation for Z_- yields the same results as (3.30) and (3.36), but with R_+ replaced by R_- in the former, with both bounded by unity.

Thus, finally, the first- and second-order statistics of the delay measurement $Z_\Delta = Z_- - Z_+$ are

$$E[Z_\Delta(\tau)] = E(Z_-) - E(Z_+) = N^2 E_c(R_-^2 - R_+^2)$$
$$= N^2 E_c[R^2(\tau - \Delta) - R^2(\tau + \Delta)], \tag{3.37}$$

$$\text{Var}[Z_\Delta(\tau)] \le \text{Var}(Z_-) + \text{Var}(Z_+) \le 2(NI_0)^2(1 + 2NE_c/I_0). \tag{3.38}$$

The reason for the last inequality is that both early and late measurements observe the same received signal over nondisjoint intervals. Thus, the correlation or covariance of Z_+ and Z_- will be positive. Since the variance of the difference equals the sum of the variances minus twice the covariance, a positive covariance diminishes the sum in (3.38).

3.5.2 Time Tracking Loop

Given the mean and variance statistics of the timing error measurement, we can model the effect of applying the measurement as a correction to the pseudorandom sequence clock generator in Figure 3.6. We define the gain as a normalized function of the relative timing error τ/T_c, in terms of (3.37),

$$E[Z_\Delta(\tau)] = N^2 E_c G(\tau/T_c),$$

where

$$G(\tau/T_c) = R^2\left(\frac{\tau - \Delta}{T_c}\right) - R^2\left(\frac{\tau + \Delta}{T_c}\right). \qquad (3.39)$$

We also define the interference variance as

$$V_0 \leq 2N^2 I_0^2(1 + 2NE_c/I_0). \qquad (3.40)$$

The gain functions $G(\ )$ for the time-limited and ideal bandlimited cases (3.3a and 3.3b) are plotted in Figure 3.7 as functions of relative timing

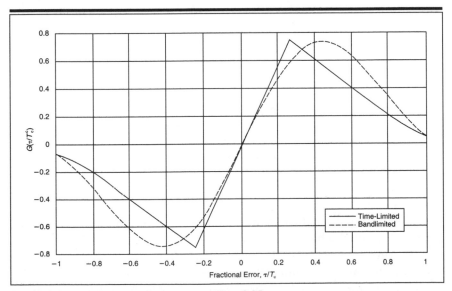

Figure 3.7 Tracking error function for $\Delta/T_c = 0.25$.

errors τ/T_c for the case $\Delta/T_c = \frac{1}{4}$. The clock timing is maintained by a voltage controlled oscillator (VCO). The clock frequency of the VCO is controlled by the time error measurement as scaled and filtered by the digital filter. The absolute timing on each chip is thus modified according to the sum of all frequency changes previously caused by timing errors. Hence, the VCO timing (phase) is modeled by an accumulator over all such previous corrections, as shown in Figure 3.8a. The scale factor α represents the gain introduced in the voltage-to-frequency conversion.

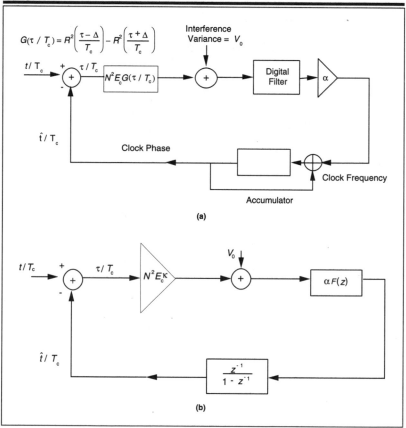

Figure 3.8 (a) Time-tracking loop model. (b) Linear model of time-tracking loop.

Once the initial synchronization has been acquired, the relative timing error will normally be small ($|\tau| \ll \Delta$). Also note from Figure 3.7 that the tracking error will usually be small enough that the error measurement can be taken as a linear function of the relative timing error τ/T_c, with scale factor equal to the slope of $G(\tau)$ at $\tau = 0$. This leads to the linear model of Figure 3.8b, with gain scale factor $N^2 E_c \kappa$, where

$$\kappa = \frac{dG(\tau/T_c)}{d(\tau/T_c)}\bigg|_{\tau=0}.$$

For the two special cases of Figure 3.7, these are

$$\kappa = \frac{d}{d(\tau/T_c)}\left[\left(1 - \left|\frac{\tau - \Delta}{T_c}\right|\right)^2 - \left(1 - \left|\frac{\tau + \Delta}{T_c}\right|\right)^2\right]\bigg|_{\tau=0} \quad (3.41)$$

$$= 4(1 - \Delta/T_c) \qquad \text{(time-limited)}$$

and

$$\kappa = \frac{d}{d(\tau/T_c)}\left\{\left[\frac{\sin[\pi(\tau - \Delta)/T_c]}{\pi(\tau - \Delta)/T_c}\right]^2 - \left[\frac{\sin[\pi(\tau + \Delta)/T_c]}{\pi(\tau + \Delta)/T_c}\right]^2\right\}\bigg|_{\tau=0}$$

$$= \frac{4\sin(\pi\Delta/T_c)}{\pi(\Delta/T_c)^2} \qquad (3.42)$$

$$\times \left[\frac{\sin(\pi\Delta/T_c)}{\pi\Delta/T_c} - \cos(\pi\Delta/T_c)\right] \qquad \text{(bandlimited)}.$$

If we take $\Delta = T_c/4$,

$$\kappa = 3 \qquad \text{(time-limited)},$$

$$\kappa = \frac{32}{\pi}\left(\frac{4}{\pi} - 1\right) = 2.78 \qquad \text{(bandlimited)}.$$

Within the linear model, the (one measurement-time) delay operator is denoted as z^{-1}, as is conventional for discrete linear system analysis. We may obtain the generating function for the timing error due to an initial relative error (τ_0/T_c), which is represented by a step function whose generating function is $(\tau_0/T_c)/(1 - z^{-1})$, as

$$\frac{\bar{\tau}(z)}{T_c} = \frac{\tau_0}{T_c}\left(\frac{1}{1 - z^{-1}}\right)\frac{1}{1 + H(z)},$$

where

$$H(z) = \frac{N^2 E_c \kappa \alpha F(z)z^{-1}}{1 - z^{-1}}. \qquad (3.43)$$

The variance of the relative timing error is obtained from the linear model as

$$\text{Var}(\tau/T_c) = \frac{V_0}{(N^2 E_c)^2 \kappa^2} \oint \frac{H(z)H(z^{-1})}{[1 + H(z)][1 + H(z^{-1})]} \frac{dz}{2\pi\, iz}. \qquad (3.44)$$

Integration is around the unit circle.

If the digital filter is omitted (except for the gain α), the tracking loop is of first order, so that (3.43) becomes

$$\frac{\tilde{\tau}(z)}{T_c} = \frac{\tau_0}{T_c} \frac{1}{1 - z^{-1}(1 - N^2 E_c \kappa \alpha)}. \qquad (3.45)$$

This means that after L measurement times, $\tau/T_c = (\tau_0/T_c)(1 - N^2 E_c \kappa \alpha)^L$, which requires $\alpha < 1/(\kappa N^2 E_c)$ for stability. Of course, the smaller the value of α, the more slowly the error decays. The variance expression of (3.44) becomes, in this case,

$$\begin{aligned}
\text{Var}(\tau/T_c) &= \oint \frac{V_0 \alpha^2\, z^{-1} dz}{2\pi i [1 - z^{-1}(1 - N^2 E_c \kappa \alpha)][1 - z(1 - N^2 E_c \kappa \alpha)]} \\
&= \frac{V_0\, \alpha^2}{1 - (1 - N^2 E_c \kappa \alpha)^2} = \frac{V_0 \alpha}{N^2 E_c \kappa (2 - N^2 E_c \kappa \alpha)},
\end{aligned}$$

where V_0 is bounded by (3.40) and $\alpha < 1/(N^2 E_c \kappa)$.

Therefore,

$$\text{Var}(\tau/T_c) < \frac{2 I_0^2 (1 + 2 N E_c/I_0) \alpha}{E_c \kappa}. \qquad (3.46)$$

This can be made arbitrarily small by choosing α sufficiently small, with a correspondingly slow decay in initial error, the consequence of a very narrow loop bandwidth.

We can achieve improved lock-in behavior, with the same variance, by using a second-order loop with a digital filter whose transfer function employs "proportional-plus-rate compensation,"

$$F(z) = 1 + \frac{az^{-1}}{1 - z^{-1}}. \qquad (3.47)$$

The resulting response to initial error and variance are given by (3.43) and (3.44).

3.6 Carrier Synchronization

This chapter dealt with the time synchronization of binary signals spread by pseudorandom sequences. We have shown that the sequence timing can be acquired and tracked accurately, and hence the spreading can be removed, without any knowledge of carrier phase and with only a rough estimate of carrier frequency. Once timing has been acquired, phase and frequency can be accurately estimated by conventional phase-lock loop techniques. Their analysis, similar to that described in Section 3.5, is well covered in many texts [Viterbi, 1966; Gardner, 1968; Lindsey and Simon, 1973; Proakis, 1989]. Even in Rayleigh fading, average phase and frequency can be accurately tracked, provided the tracking loop bandwidth is narrow compared to the fading bandwidth.

Likelihood Functions and Probability Expressions

3.A.1 Bayes and Neyman-Pearson Hypothesis Testing

Acquiring synchronization and digital communication both involve testing binary hypotheses: whether a signal is present or absent in the first case, or whether a zero or a one was transmitted in the second case. The model assumes that a number N of observations or measurements have been made on the signal, defined as the vector $\mathbf{y} = y_1, y_2, \ldots, y_N$. It also assumes that conditional probabilities are known for the observations under the two hypotheses, which we denote "0" and "1." The binary symbols refer to signal absent and present in the synchronization case, and to transmission of "0" or "1" in the binary communication case. We denote these conditional probabilities by $p_0(\mathbf{y})$ and $p_1(\mathbf{y})$, usually called likelihood functions. $\mathbf{y}$ is generally a vector of real numbers, although we can also treat the case where the components of $\mathbf{y}$ are from a finite nonnumerical set.

The Bayes criterion [Helstrom, 1968] seeks to minimize the "cost" of an error. However, it requires that we know the a priori probabilities of "0" and "1," as well as the set of costs of choosing "i" when "j" is true, where "i" and "j" can be either "0" or "1." The Neyman–Pearson criterion [Helstrom, 1968; Neyman and Pearson, 1933] requires neither of the above. It seeks to minimize the probability of choosing "0" when "1" is true (a "miss") for a given acceptable probability of choosing a "1" when a "0" is true (a "false alarm").

Applying either criterion gives rise to the likelihood ratio test:

$$\Lambda(\mathbf{y}) \triangleq \frac{p_a(\mathbf{y})}{p_b(\mathbf{y})} \gtrless \psi. \qquad (3A.1)$$

Here, $a = 0$ or 1 and $b = \bar{a}$, and ψ is a numerical threshold that depends on the optimized parameters of either test. If the ratio exceeds ψ, hypothesis "a" is chosen, while if it does not, hypothesis "b" is chosen. For binary communication, the Bayes rule, with equal a priori probabilities and equal costs of mistaking "a" for "b" or vice versa, requires $\psi = 1$. For synchronization, ψ is chosen to fulfill the requirement that the false alarm probability not exceed a given value.

When the observations $y_1, y_2, \ldots, y_N$ are mutually independent, as in a memoryless channel, the likelihood ratio can be written as a product and its logarithm as a sum. Then the test reduces to

$$\ln \Lambda(\mathbf{y}) = \sum_{j=1}^{N} \ln[p_a(y_j)/p_b(y_j)] \gtrless \theta. \tag{3A.2}$$

Here, $\theta = \ln \Psi$, which equals 0 for the equiprobable, equal cost, binary communication case.

3.A.2 Coherent Reception in Additive White Gaussian Noise

The simplest application of the log likelihood test, (3A.2), is coherent reception of a fixed-amplitude, known-phase signal in additive white Gaussian noise of one-sided density, I_0. We assume that $a = $ "0" is represented (prior to spreading by PN multiplication) as N chips of amplitude $+\sqrt{E_c}$ and $b = $ "1" is represented as N chips of amplitude $-\sqrt{E_c}$. Then if $\mathbf{y}$ is the vector of outputs (after despreading by multiplication by the same PN sequence),

$$p_0(\mathbf{y}) = \prod_{n=1}^{N} e^{-(y_a - \sqrt{E_c})^2/I_0}/\sqrt{\pi I_0},$$

$$p_1(\mathbf{y}) = \prod_{n=1}^{N} e^{-(y_a + \sqrt{E_c})^2/I_0}/\sqrt{\pi I_0}. \tag{3A.3}$$

Applying (3A.2), we obtain

$$\ln \Lambda(\mathbf{y}) = \frac{4\sqrt{E_c}}{I_0} \sum_{n=1}^{N} y_n \gtrless \theta. \tag{3A.4}$$

For binary communication with equal a priori probabilities and costs, $\theta = 0$. Thus, we may ignore the scale factor in (3A.4) and choose a "0" if the sum is positive and a "1" if it is negative. We then do not need to know the energy-to-noise ratio.

3.A.3 Noncoherent Reception in AWGN for Unfaded Signals

Suppose now that the amplitude is fixed, but that the phase is unknown and taken to be a uniformly distributed random variable. For the synchronization case, with "1" and "0" denoting signal present and absent, respectively, the observable vectors for in-phase and quadrature outputs may be taken as

$$\mathbf{y}^{(I)} = (y_1^{(I)}, y_2^{(I)}, \ldots, y_N^{(I)}) \quad \text{and} \quad \mathbf{y}^{(Q)} = (y_1^{(Q)}, y_2^{(Q)}, \ldots, y_N^{(Q)}).$$

Then, for reception with random phase, ϕ, in AWGN with variance $I_0/2$, we obtain the likelihood functions in terms of both sets of observables and conditioned on ϕ. When signal is present,

$$p_1(\mathbf{y}^{(I)}, \mathbf{y}^{(Q)} \mid \phi) = \prod_{n=1}^{N} \exp[-(y_n^{(I)} - \sqrt{E_c} \cos \phi)^2/I_0]$$

$$\exp[-(y_n^{(Q)} - \sqrt{E_c} \sin \phi)^2/I_0]/\pi I_0.$$

When signal is not present,

$$p_0(\mathbf{y}^{(I)}, \mathbf{y}^{(Q)}) = \prod_{n=1}^{N} \exp[-(y_n^{(I)})^2/I_0] \exp[-(y_n^{(Q)})^2/I_0]/\pi I_0. \quad (3A.5)$$

Since the ratio of the two functions depends only on the sums

$$Y^{(I)} = \sum_{n=1}^{N} y_n^{(I)}, \qquad Y^{(Q)} = \sum_{n=1}^{N} y_n^{(Q)}, \qquad (3A.6)$$

we may take the sums to be the observables. The variances of $Y^{(I)}$ and $Y^{(Q)}$ are $V/2$, where

$$V \triangleq NI_0, \qquad (3A.7)$$

while their means[4] are $N\sqrt{E_c} \cos \phi$ and $N\sqrt{E_c} \sin \phi$, respectively. We thus

---

[4] Throughout this appendix, we assume exact timing so that $R(\tau) = 1$. When this is not the case, E_c must be everywhere multiplied by $R^2(\tau)$, as has been done throughout Chapter 3.

may express the likelihood functions of $Y^{(I)}$ and $Y^{(Q)}$ as

$$p_1(Y^{(I)}, Y^{(Q)} \mid \phi) = \exp\{-[Y^{(I)} - N\sqrt{E_c}\cos\phi]^2/V\}$$
$$\times \exp\{-[Y^{(Q)} - N\sqrt{E_c}\sin\phi]^2/V\}/\pi V, \quad (3A.8)$$

$$p_0(Y^{(I)}, Y^{(Q)}) = \exp\{-[(Y^{(I)})^2 + (Y^{(Q)})^2]/V\}/\pi V. \quad (3A.9)$$

Since the phase ϕ is a uniform random variable, we may obtain the unconditional likelihood function, when signal is present, by averaging (3A.8) over ϕ to obtain

$$p_1(Y^{(I)}, Y^{(Q)}) = \int_0^{2\pi} p(Y^{(I)}, Y^{(Q)} \mid \phi)\, d\phi/2\pi$$

$$= \int_0^{2\pi} \exp\{-[(Y^{(I)})^2 + (Y^{(Q)})^2 + N^2 E_c]/V\}$$

$$\times \frac{\exp\{2N\sqrt{E_c}[Y^{(I)}\cos\phi + Y^{(Q)}\sin\phi]/V\}}{\pi V} \frac{d\phi}{2\pi}$$

$$= \frac{1}{\pi V} \exp\left(\frac{-[(Y^{(I)})^2 + (Y^{(Q)})^2]}{V}\right) \quad (3A.10)$$

$$\times \mathcal{I}_0\left(\frac{2\sqrt{N^2 E_c[(Y^{(I)})^2 + (Y^{(Q)})^2]}}{V}\right) \exp(-N^2 E_c/V).$$

where,

$$\mathcal{I}_0(x) \triangleq \int_0^{2\pi} \exp(x\cos\phi)\, d\phi/2\pi$$

is the zeroth-order modified Bessel function.

Finally, defining the output variable

$$Z \triangleq Y^{(I)2} + Y^{(Q)2} \quad (3A.11)$$

and the squared mean[5]

$$M^2 = N^2 E_c, \quad (3A.12)$$

---

[5] In (3.9) *et seq.* of Section 3.3 we replace M by M_D, which is the mean reduced by the scale factor $R(\tau)$.

we obtain by proper transformation of (3A.9) and (3A.10),

$$p_0(Z) = \frac{\exp(-Z/V)}{V}, \tag{3A.13}$$

$$p_1(Z) = \frac{\exp[-(Z + M^2)/V]}{V} \, \mathcal{I}_0\left(\frac{2\sqrt{M^2Z}}{V}\right). \tag{3A.14}$$

We note also that if we normalize Z by V, letting $z = Z/V$ and $\mu = M^2/V$, we have the likelihood functions of the normalized variables,

$$p_0(z) = e^{-z}, \tag{3A.13a}$$

$$p_1(z) = e^{-(z+\mu)}\mathcal{I}_0(2\sqrt{\mu z}). \tag{3A.14a}$$

The log likelihood ratio test becomes

$$\ln \Lambda(Z) = \ln[p_1(Z)/p_0(Z)] = \ln \mathcal{I}_0(2\sqrt{M^2Z}/V) - M^2/V \gtrless \theta. \tag{3A.15}$$

However, since $\mathcal{I}_0(c\sqrt{Z})$ is monotonic in Z for all positive Z, the test may be simplified to

$$Z/V = z \gtrless \hat{\theta}. \tag{3A.16}$$

$\hat{\theta}$ is obtained from the Neyman–Pearson constraint that the *false alarm probability be kept at the value* α. Then using (3A.13a), we have

$$\alpha = \Pr(z > \hat{\theta} \mid \text{no signal}) = \int_{\hat{\theta}}^{\infty} p_0(z)\, dz$$
$$= e^{-\hat{\theta}},$$

so that

$$\hat{\theta} = -\ln \alpha. \tag{3A.17}$$

3.A.4 Noncoherent Reception of Multiple Independent Observations of Unfaded Signals in AWGN

Suppose that, as in Figure 3.1a, we make L independent observations (successively or on different multipath elements, as in Chapter 4) whose unknown phases are independent random variables. From (3A.15), the log likelihood function of the L observations becomes

$$\ln \Lambda(Z_1, Z_2, \ldots, Z_L) = \sum_{\ell=1}^{L} \ln[p_1(Z_\ell)/p_0(Z_\ell)]$$

$$= \sum_{\ell=1}^{L} [\ln \mathcal{I}_0(2\sqrt{M^2 Z_\ell}/V) - M^2/V] \gtrless \theta. \qquad (3\text{A}.18)$$

This requires knowledge of both E_c and V, which may not be available or easily measurable. On the other hand, for small arguments, $\ln \mathcal{I}_0(x) \sim x^2/4$, so that a reasonable approximation of the test is to simply sum the Z_ℓ variables, resulting in the suboptimal test,

$$Z \triangleq \sum_{\ell=1}^{L} Z_\ell \gtrless \tilde{\theta}. \qquad (3\text{A}.19)$$

Since Z is then the decision variable, we require its likelihood function under both hypotheses. Since it is the sum of L independent variables, each of which has a likelihood function given by (3A.13) or (3A.14), we determine the characteristic function (Laplace transform) of each, take its Lth power, and invert the transform to obtain the desired probability density (likelihood) function. For the "0" hypothesis, we have from (3A.13)

$$M_0(s) = \int_0^\infty p_0(Z) e^{-sZ}\, dZ = \frac{1}{1 + Vs}. \qquad (3\text{A}.20)$$

Then, for the sum of L variables,

$$p_0(Z) = \oint M_0^L(s) e^{sZ}\, ds/(2\pi i)$$

$$= \oint \frac{e^{sZ}}{(1 + Vs)^L} \frac{ds}{2\pi i} \qquad (3\text{A}.21)$$

$$= \frac{Z^{L-1} e^{-Z/V}}{(L-1)! V^L}.$$

Similarly, from (3A.14), we obtain

$$M_1(s) = \int_0^\infty p_1(Z) e^{-sZ}\, dZ$$

$$= \frac{\exp[-M^2 s/(1 + Vs)]}{1 + Vs}. \qquad (3\text{A}.22)$$

Consequently, for the sum of L variables,

$$
\begin{aligned}
p_1(Z) &= \oint M_1^L(s)e^{sZ}\, ds/(2\pi i) \\
&= \oint \frac{\exp\{s[Z - LM^2/(1 + Vs)]\}}{(1 + Vs)^L}\, \frac{ds}{2\pi i} \qquad\qquad (3A.23) \\
&= \frac{1}{V}\left(\frac{Z}{LM^2}\right)^{(L-1)/2} \exp\left(\frac{-Z - LM^2}{V}\right) \mathcal{I}_{L-1}\left(\frac{2\sqrt{LM^2Z}}{V}\right).
\end{aligned}
$$

For the normalized variable $z = Z/V$ with $\mu = M^2/V$, we have the likelihood functions

$$
p_0(z) = \frac{z^{L-1}e^{-z}}{(L-1)!}, \qquad\qquad (3A.21a)
$$

$$
p_1(z) = \left(\frac{z}{L\mu}\right)^{(L-1)/2} e^{-(z + L\mu)}\mathcal{I}_{L-1}(2\sqrt{L\mu z}). \qquad (3A.23a)
$$

3.A.5 Noncoherent Reception of Rayleigh-Faded Signals in AWGN

Suppose now that the amplitude is Rayleigh-distributed. Then the chip energy E_c is multiplied by α^2 where α is Rayleigh-distributed. Hence,

$$
p(\alpha) = \frac{2\alpha}{\sigma^2}e^{-\alpha^2/\sigma^2}. \qquad\qquad (3A.24)
$$

Letting $\beta = \alpha^2$,

$$
p(\beta) = \frac{e^{-\beta/\sigma^2}}{\sigma^2}, \qquad\qquad (3A.25)
$$

and the average chip energy becomes

$$
\bar{E}_c = \overline{\alpha^2}E_c = \bar{\beta}E_c = \sigma^2 E_c. \qquad\qquad (3A.26)
$$

Then, for $L = 1$, $p_0(Z)$ is still given by (3A.13) since it does not depend on the (fading) signal power. However, $p_1(Z)$ of (3A.14) is modified by weighing the signal power, or equivalently M^2, by β and averaging. Then

$$p_1(Z) = \int_0^\infty \frac{e^{-\beta/\sigma^2}}{\sigma^2} \frac{\exp[(-Z + \beta M^2)/V]}{V} \mathcal{I}_0\left(\frac{2\sqrt{\beta M^2 Z}}{V}\right) d\beta$$

$$= \frac{\exp[-Z/(V + M^2\sigma^2)]}{V + M^2\sigma^2} \qquad (3A.27)$$

$$\triangleq \frac{\exp[-Z/(V + \overline{M}^2)]}{V + \overline{M}^2},$$

where $\overline{M}^2 \triangleq M^2\sigma^2 = N^2\overline{E}_c$.

Thus, defining

$$V_F = V + \overline{M}^2 = N(I_0 + N\overline{E}_c), \qquad (3A.28)$$

$$p_0(Z) = \frac{1}{V} e^{-Z/V}, \qquad (3A.29)$$

$$p_1(Z) = \frac{1}{V_F} e^{-Z/V_F}.$$

Similarly, for $L > 1$, Z is the sum of the squares of L independent Rayleigh observations. Hence, it follows from (3A.21) that

$$p_0(Z) = \frac{Z^{L-1} e^{-Z/V}}{(L-1)! V^L} \qquad (3A.30)$$

and

$$p_1(Z) = \frac{Z^{L-1} e^{-Z/V_F}}{(L-1)! V_F^L}, \qquad (3A.31)$$

$$\text{where } V_F = V + \overline{M}^2. \qquad (3A.32)$$

For the normalized variable $z = Z/V$, with $\overline{\mu} = \overline{M}^2/V$ and $V_F = V(1 + \overline{\mu})$, the likelihood functions become

$$p_0(z) = \frac{z^{L-1} e^{-z}}{(L-1)!}, \qquad (3A.30a)$$

$$p_1(z) = \frac{z^{L-1} e^{-z/(1+\overline{\mu})}}{(L-1)! (1 + \overline{\mu})^L}. \qquad (3A.31a)$$

Modulation and Demodulation of Spread Spectrum Signals in Multipath and Multiple Access Interference

4.1 Purpose

Throughout the last two chapters, we have approximated the interference to a given user from all other multiple access users by a Gaussian process. We have also assumed in most cases that the desired user's signal was constant in amplitude, not subject to fading due to multipath propagation or blockage. We now reexamine both constraints. In the next section, we show that the Gaussian interference assumption is often unnecessary if performance is evaluated using relatively tight upper bounds. The rest of the chapter deals with modulation waveform and demodulation alternatives, and their resulting performance with multipath propagation and multiple access users. The goal is to achieve acceptable performance with the lowest possible bit-energy-to-interference, E_b/I_0, so as to maximize capacity, according to (1.5).

4.2 Chernoff and Bhattacharyya Bounds

We begin by evaluating some well-known bounding techniques for constant amplitude and phase signals, coherently demodulated in additive Gaussian noise. We then proceed to show that the same bounds hold in some cases when the actual multiple access interference distributions are used in place of the Gaussian approximation. These bounding techniques will be used for a variety of purposes throughout the remainder of this text.

The Chernoff bound [Gallager, 1968] on a distribution is an extension of the older Chebyshev bound. It is obtained by upper-bounding the unit step function inherent in the following calculation. For the random variable x, we can express the complementary distribution function

$$\Pr(x > X) = \int_{-\infty}^{\infty} u(x - X)\, dF(x) = E[u(x - X)],$$

where $u(\ )$ is the unit step function,

$$u(\xi) = \begin{cases} 1 & \text{if } \xi \geq 0, \\ 0 & \text{otherwise,} \end{cases} \tag{4.1}$$

and $F(x)$ is the distribution function.

Then since

$$u(\xi) \leq \exp(\rho\xi), \qquad \rho > 0, \tag{4.2}$$

with equality if and only if $\xi = 0$, it follows that

$$\Pr(x > X) < E\{\exp[\rho(x - X)]\}$$
$$= e^{-\rho X} \int_{-\infty}^{\infty} e^{\rho x}\, dF(x), \qquad \rho > 0. \tag{4.3}$$

Minimizing with respect to ρ, we obtain

$$\Pr(x > X) < \underset{\rho > 0}{\text{Min}}\, e^{-\rho X} E(e^{\rho x}), \tag{4.4}$$

which is the *Chernoff bound*. In most cases this bound can be reduced by a factor of 2, as will be shown in Section 4.2.4.

When the random variable represents a likelihood function and $X = 0$, so that the probability (4.3) becomes an error probability, applying the Chernoff bound gives rise to another well-known bound. Let y be the decision variable, and let $p_0(y)$ and $p_1(y)$ be the respective probability density functions under the hypotheses that "0" or "1" was sent. Then given that "0" was sent,

$$P_{E_0} = \Pr\left(\ln \frac{p_1(y)}{p_0(y)} > 0 \,\Big|\, \text{"0" sent}\right)$$
$$< E[\exp\{\rho \ln [p_1(y)/p_0(y)]\} \mid \text{"0" sent}]$$
$$= \int_{-\infty}^{\infty} \left[\frac{p_1(y)}{p_0(y)}\right]^\rho p_0(y)\, dy = \int_{-\infty}^{\infty} p_0^{1-\rho}(y)p_1^\rho(y)\, dy, \qquad \rho > 0.$$

The same applies to P_{E_1} with $p_0(y)$ and $p_1(y)$ interchanged. Choosing $\rho = \frac{1}{2}$ leads to the particularly simple expression[1]

$$P_E \le \int_{-\infty}^{\infty} \sqrt{p_0(y)p_1(y)}\, dy, \qquad (4.5)$$

known as the *Bhattacharyya bound*.

4.2.1 Bounds for Gaussian Noise Channel

As an example application of both the Chernoff bound (4.4) and the Bhattacharyya bound (4.5) and for later comparison, consider coherent reception in Gaussian noise with energy-to-noise density E/I_0. Then the normalized demodulator output variable y is Gaussian, with variance $I_0/2$ and mean $\pm\sqrt{E}$. The sign depends on whether the transmitted symbol had a positive or negative sign corresponding to "0" or "1," respectively. Then, since by symmetry $P_{E_0} = P_{E_1} = P_E$,

$$P_E = \Pr(y > 0 \mid -) = \int_0^\infty \exp[-(y + \sqrt{E})^2/I_0]\, dy/\sqrt{\pi I_0}$$
$$= \int_{\sqrt{2E/I_0}}^{\infty} e^{-x^2/2}\, dx/\sqrt{2\pi} = Q(\sqrt{2E/I_0}). \qquad (4.6)$$

The *Chernoff* bound to this probability is obtained using (4.4), where x has mean $-\sqrt{E}$ and variance $I_0/2$, and $X = 0$. Thus,

$$E(e^{\rho x}) = \int_{-\infty}^{\infty} e^{\rho x} e^{-(x+\sqrt{E})^2/I_0}\, dx/\sqrt{\pi I_0}$$
$$= \exp[(\rho I_0 - 2\sqrt{E})^2/4I_0]e^{-E/I_0}, \qquad \rho > 0, \qquad (4.7)$$

---

[1] When $p_0(y)$ and $p_1(y)$ satisfy appropriate symmetry conditions and 0 and 1 are a priori equally probable, this choice minimizes the bound. This is the case for the example in the next subsection.

and

$$P_E < \operatorname*{Min}_{\rho>0} E(e^{\rho x}). \tag{4.8}$$

This expression is minimized by the choice $\rho = 2\sqrt{E}/I_0$, for which

$$P_E < e^{-E/I_0}. \tag{4.9}$$

The same result is obtained from the Bhattacharyya bound, as can be seen by applying (4.5) with

$$
\begin{aligned}
p_0(y) &= \exp[-(y - \sqrt{E})^2/I_0]/\sqrt{\pi I_0}, \\
p_1(y) &= \exp[-(y + \sqrt{E})^2/I_0]/\sqrt{\pi I_0}.
\end{aligned}
\tag{4.10}
$$

Hence,

$$P_E < \int_{-\infty}^{\infty} \sqrt{p_0(y)p_1(y)}\, dy = e^{-E/I_0} \int_{-\infty}^{\infty} e^{-y^2/I_0}\, dy/\sqrt{\pi I_0} = e^{-E/I_0}. \tag{4.11}$$

We now show that for time-synchronous multiple access interference, possibly in addition to background Gaussian noise, the Chernoff bound yields this same result.

4.2.2 Chernoff Bound for Time-Synchronous Multiple Access Interference with BPSK Spreading

Consider first BPSK spread spectrum CDMA, with a random phase ϕ but where all k_u users are received with identical chip timing. Then the output of the summation device of Figure 2.2 is

$$Y = \sum_{n=1}^{N} y_n.$$

Here, ignoring interchip interference,

$$y_n = \sqrt{E_c(k)}\, x_n(k) + \sum_{j \neq k} \sqrt{E_c(j)}\, z_n(j) + v_n, \tag{4.12}$$

and $z_n(j)$ are the other-user chip outputs (including the product of different spreading sequences). v_n is the effect of background Gaussian noise on the output. Note that because of the time synchronization[2], the other users can no longer be treated as random processes, but rather as a sequence of binary random variables with random phase shift.

Thus, with $x_n(k) = -1$ for all n,

$$P_E = \Pr(Y > 0 \mid -) < E\left\{ \exp\left[\rho \sum_{n=1}^{N} y_n \right] \right\}$$

$$< \exp[-\rho N \sqrt{E_c(k)}] \prod_{n=1}^{N} \left[E[\exp(\rho \nu_n)] \prod_{j=1}^{k_u} E\{\exp[\rho \sqrt{E_c(j)}\, z_n(j)]\} \right],$$

$$\rho > 0. \quad (4.13)$$

The last step follows from the independence of all interfering users and of successive chips from each user.

Since ν_n is Gaussian with zero mean and variance $N_0/2$, and letting $E = 0$ and $I_0 = N_0$ in (4.7), we have

$$E[e^{\rho \nu_a}] = \exp(N_0 \rho^2/4). \quad (4.14)$$

We see from Figure 2.2 that each interfering user chip has amplitude $\pm \sqrt{E_c(j)}$ and phase

$$\phi_j' \triangleq \phi_j - \phi_k, \qquad j \neq k.$$

Both are random variables, the first binomial and the second uniform. Thus,

$$E[e^{\rho \sqrt{E_c(j)}\, z_n(j)} \mid \phi_j'] = \tfrac{1}{2}\{\exp[\sqrt{E_c(j)}(\cos \phi_j')\rho] + \exp[-\sqrt{E_c(j)}(\cos \phi_j')\rho]\}.$$

But since ϕ_j' is a uniformly distributed random variable on $(0, 2\pi)$,

$$E[e^{\rho \sqrt{E_c(j)}\, z_n(j)}] = \mathcal{I}_0(\sqrt{E_c(j)}\rho) < \exp[\rho^2 E_c(j)/4], \quad (4.15)$$

since the zeroth-order Bessel function $\mathcal{I}_0(x) < e^{x^2/4}$. Thus, combining (4.13) through (4.15), we obtain

$$P_E < \operatorname*{Min}_{\rho > 0}\, \exp\{-N[\rho \sqrt{E_c(k)} - \rho^2 I_0/4]\}, \quad (4.16)$$

---

[2] In terms of a cellular network, this is a proper model for multiple users of the forward link from the same base station, or of multiple users of the reverse link whose received timing has been synchronized.

where

$$I_0 = N_0 + \sum_{j=1}^{k_u} E_c(j).$$

Hence, with the minimizing value of $\rho = 2\sqrt{E_c(k)}/I_0$,

$$P_E < \exp[-NE_c(k)/I_0], \tag{4.17}$$

exactly as for a Gaussian approximation (4.9). However, there is a problem with this result. When the number of users k_u is large it is reasonable to take the expectation with respect to phase difference ϕ_j. But for small k_u (e.g., a single interfering user or base station) this may not be justifiable. We remedy this, as before, by using QPSK spreading.

4.2.3 Chernoff Bound for Time-Synchronous Multiple Access Interference with QPSK Spreading

We proceed as for BPSK spreading, but, in place of (4.12) we have, from Figure 2.4,

$$y_n = \sqrt{E_c(k)}\, x_n(k) + \sum_{j \neq k} [z_n^{(I)}(j) + z_n^{(Q)}(j)] + \nu_n. \tag{4.18}$$

Here, the random variable $z_n^{(I)}(j)$ takes on the value $\sqrt{E_c(j)/2}\, \hat{z}_n^{(I)}(j) \cos \phi_j'$ and the random variable $z_n^{(Q)}(j)$ takes on the value $\sqrt{E_c(j)/2}\, \hat{z}_n^{(Q)}(j) \sin \phi_j'$, where $\hat{z}_n^{(I)}(j)$ and $\hat{z}_n^{(Q)}(j)$ are independent binary random variables and $\phi_j' = \phi_j - \phi_k - \pi/4$. Thus, we have in place of (4.13)

$$P_E < \exp[-\rho N \sqrt{E_c(k)}] \prod_{n=1}^{N}$$

$$\left[E[\exp(\rho \nu_n)] \prod_{j=1}^{k_u} E\{\exp[\rho \sqrt{E_c(j)/2}\, \hat{z}_n^{(I)}(j) \cos \phi_j'] \right. \tag{4.19}$$

$$\left. \exp[\rho \sqrt{E_c(j)/2}\, \hat{z}_n^{(Q)}(j) \sin \phi_j'] \} \right].$$

The expectation of the terms in brackets that now replace (4.15) becomes

$$
\begin{aligned}
E\{&\exp[\rho\sqrt{E_c(j)}/2\ \hat{z}_n^{(I)}(j)\cos\phi_j']\exp[\rho\sqrt{E_c(j)}/2\ \hat{z}_n^{(Q)}(j)\sin\phi_j']\} \\
&= \tfrac{1}{2}\,[\exp(\rho\sqrt{E_c(j)}/2\cos\phi_j') + \exp(-\rho\sqrt{E_c(j)}/2\cos\phi_j')] \\
&\quad \times \tfrac{1}{2}\,[\exp(\rho\sqrt{E_c(j)}/2\sin\phi_j') + \exp(-\rho\sqrt{E_c(j)}/2\sin\phi_j')] \qquad (4.20)\\
&= \cosh(\rho\sqrt{E_c(j)}/2\cos\phi_j')\cosh(\rho\sqrt{E_c(j)}/2\sin\phi_j') \\
&< \exp[\rho^2 E_c(j)(\cos^2\phi_j')/4]\exp[\rho^2 E_c(j)(\sin^2\phi_j')/4] \\
&= \exp[\rho^2 E_c(j)/4],
\end{aligned}
$$

which follows from the inequality

$$
\cosh(x) = (e^x + e^{-x})/2 = \sum_{k\ \text{even}} x^k/k! < e^{x^2/2}. \qquad (4.21)
$$

Thus, (4.20) is the same bound as (4.15) for the BPSK case. Using (4.20) and (4.14) in (4.19) leads to (4.16), just as for the BPSK case, which again yields the same final result (4.17).

The difference is that in (4.20), for the QPSK case, ϕ_j' can take on any value and does not need to be regarded as a random variable. In fact, we can even take $\phi_j' = 0$, allowing, for example, all users (from a common base station) to employ the same phase. The important conclusion is that the Chernoff bound (4.17) holds equally for time-synchronous multiple access interference and for Gaussian noise. Thus, the central limit approximation is no longer required. For time-asynchronous reception (usually the case for the reverse link), this argument does not apply. However, random chip timing can only make the central limit approximation more appropriate. Furthermore, if the Chernoff bound is the same as for Gaussian interference even without the randomness of chip timing, it stands to reason that the randomness cannot significantly alter this situation.

4.2.4 Improving the Chernoff Bound by a Factor of 2

In most cases of interest, the probability whose bound we seek is the integral of the tail of a probability density function $f(x)$. Thus, as in (4.6), it can be expressed as

$$
\Pr(Y > 0) = \int_0^\infty f(y)\, dy,
$$

where $E(Y) < 0$.

Often in such cases we can show that

$$f(y) \le f(-y) \qquad \text{for all } y > 0. \tag{4.22}$$

This is always the case if $f(y)$ is symmetric about its mean and monotonically decreasing with respect to $|y - E(Y)|$. Then the moment generating function

$$
\begin{aligned}
E(e^{\rho Y}) &= \int_{-\infty}^{\infty} e^{\rho y} f(y)\, dy = \int_{-\infty}^{0} e^{\rho y} f(y)\, dy + \int_{0}^{\infty} e^{\rho y} f(y)\, dy \\
&= \int_{0}^{\infty} e^{-\rho y} f(-y)\, dy + \int_{0}^{\infty} e^{\rho y} f(y)\, dy \\
&\ge \int_{0}^{\infty} (e^{-\rho y} + e^{\rho y}) f(y)\, dy = 2 \int_{0}^{\infty} \cosh(\rho y) f(y)\, dy \\
&\ge 2 \int_{0}^{\infty} f(y)\, dy = 2\Pr(Y > 0).
\end{aligned}
$$

Thus,

$$\Pr(Y > 0) \le \tfrac{1}{2} E(e^{\rho Y}) \qquad \text{for all } \rho, \tag{4.23}$$

provided condition (4.22) is satisfied. In cases where Y is the sum of a large number of independent variables, this is often the case.

4.3 Multipath Propagation: Signal Structure and Exploitation

In terrestrial communication, and to a lesser extent in communication with satellites at low angular elevations, the transmitted signal is reflected and refracted by a variety of smooth and rough terrains, so that it is replicated at the receiver with several time delays. This is called multipath propagation. Each individual path also arrives at its own amplitude and carrier phase. Propagation characteristics are qualitatively the same for all forms of signal structure, though they will vary quantitatively with carrier frequency and terrain characteristics [Schwartz *et al.*, 1966; Jakes, 1974; Turin, 1980; Lee, 1989]. The structures of the individual propagation paths can be identified and possibly exploited only to the extent that they can be distinguished from one another. The wider the instantaneous bandwidth of the signals, the more distinguishable they are, since narrowband filtering tends to blur the multipath components together. In particular suppose the spread spectrum signals employ pseudorandom sequences with chip time T_c inversely proportional to the spreading bandwidth. In this case, the individual paths can be distinguished if they are mutually separated by delays greater than T_c, for then the various delayed versions of the signal will be mutually nearly uncorrelated, according to the delay-and-add property (2.10).[3]

The transmitted signal for the kth user for free space propagation would be received with amplitude $\sqrt{E_c(k)}$ at a constant phase and a unique delay. For QPSK spreading, it is instead received as the multipath signal

$$X_k(t) = \sqrt{E_c(k)} \sum_n x_n(k) \sum_{\ell=1}^{L} \alpha_\ell h(t - nT_c - \delta_\ell)$$
$$\times \{a_n^{(I)}(k) \cos[2\pi f_0(t - \delta_\ell) + \phi_\ell] \qquad (4.24)$$
$$+ a_n^{(Q)}(k) \sin[2\pi f_0(t - \delta_\ell) + \phi_\ell]\}.$$

This is shown in Figure 4.1. The path amplitudes α_ℓ will of course depend on the relative propagation distances and the reflective or refractive properties of the terrain or buildings. However, in many cases, particularly in confined areas, each of the distinguishable multipath components (i.e., those separated by more than T_c from one another) will actually be itself the linear combination of several indistinguishable paths of varying amplitudes. Since these will add as random vectors, the amplitude of each term will appear to be Rayleigh-distributed, and the phase uniformly distributed. This is the most commonly accepted model. Occasionally, however, because of specular reflection or free space propagation of a component, a path will appear as the sum of a constant amplitude component plus a Rayleigh-distributed amplitude component (the combination of several smaller indistinguishable components). This composite component then has a Rician distribution. It is most common in terrestrial propagation in urban areas with a single strong reflection, and in satellite reception of a free space component plus several reflections off close-by terrain.

To exploit energy in the multiple components of multipath propagation, they must be identified and acquired. It is particularly important to determine the relative delays and, subsequently, when possible, their amplitudes and phases. This can be performed even with fully modulated

---

[3] For the time-limited waveform, correlation beyond one chip time is zero [see (2.21) and (2.25)]. For other cases, it decreases rapidly beyond one chip time.

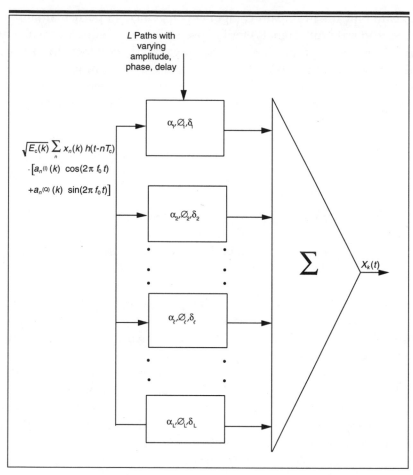

Figure 4.1 Multipath propagation model.

signals, but the estimate is much more precise and the resulting performance is much improved if the path identification and parameter estimation are performed on an unmodulated signal. Unmodulated segments
can be inserted every so often in a modulated signal, particularly with
time division multiplexing. However, in a spread spectrum system it is
much more effective and easier to separate the unmodulated "pilot" signal from the data modulated signal by assigning it an individual pseudorandom sequence.[4]

---

[4] As further elaborated in the next section, a pilot signal is well justified on the forward link
where it may be shared by the many signals transmitted from the base station to multiple
users. However, it might even be used by an individual reverse link user, if performance
improvement through (partial) coherence can justify the overhead energy required (see also
Appendix 4A).

The pilot sequence can be searched by the hypothesis testing device of Figure 3.1 with time (delay) hypotheses separated by a fraction of a chip T_c. However, once the first strong component is found, the entire search window for all components can be limited to typically a few tens of chip times, representing the total delay dispersion of the multipath propagation. The pilot pseudorandom sequence is unmodulated, and the carrier frequency is assumed to be accurately tracked. Thus, the number of chips used for estimation can be made as large as desired, limited only by the rate of change of amplitude and phase. Also, unlike the last chapter where the initial acquisition search ended when the correct hypothesis was detected, the search here will not end. Once a path is detected and verified, the search continues indefinitely, since new multipath components will appear and old ones disappear frequently, particularly for users in motion. Once found, the component sequence timing must be tracked by an early–late gate, both to refine the time estimate and to adjust for distance and velocity of users in motion.

We will explore this mechanization more fully. However, we first note that whereas the multipath propagation model of Figure 4.1 has long been accepted in the communication theory literature, it has been traditionally associated with a static receiver, in which it was assumed that each of the components (and even their number, L) remained at a constant delay, although their amplitudes and phases might vary. Employing a receiver that takes a static number of paths, in place of the varying number assumed here, complicates the form of the optimum receiver: L must generally be taken to be larger than if only the currently active path delays are being demodulated. Also, with a static receiver, motion causes a transition from one delay path to another with possible discontinuity of amplitude and phase.

In the next section we explore the optimum pilot-aided demodulator based on tracking each multipath component. In the following subsection, we consider its performance.

4.4 Pilot-Aided Coherent Multipath Demodulation

A pilot sequence for determining multipath component characteristics is well justified for one-to-many transmission channels, such as the forward (down-) link from a base station to multiple users. This is because the same pilot sequence is shared among the k_u users controlled by that base station. For the same reason, the energy devoted to the pilot can be greater than that devoted to the individual users. Thus, if $X_k(t)$ given in (4.24) is the received signal for the kth user, let $X_0(t)$ be the unmodulated pilot

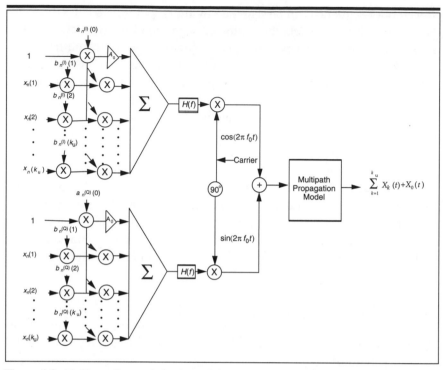

Figure 4.2 Multiuser (base station) modulator and multipath channel.

signal, so that $x_n(0) = 1$ for all n. Also, we assume that the pilot's pseudo-random sequence is shared by all users by multiplying the pilot pseudo-random (± 1) sequence with all the user-specific sequences (Figure 4.2). Hence, the received signal containing k_u users and a pilot sequence, all originating from a common base station, will be

$$X_0(t) + \sum_{k=1}^{k_u} X_k(t).$$

Here, $X_k(t)$ is given by (4.24) for $k = 1, 2, \ldots, k_u$, but $X_0(t)$ is further scaled by A_0, the additional gain allotted to the pilot signal.

As shown in Figure 4.2, $a_n^{(I)}(0)$ and $a_n^{(Q)}(0)$ are the pilot's QPSK spreading sequence. The users' sequences are the product[5] of those of the pilot

---

[5] The reason for multiplying the user sequence by the pilot sequence is to identify the user with the particular base station that is handling the call.

and of the user-specific sequences $b_n^{(I)}(k)$ and $b_n^{(Q)}(k)$. That is,

$$a_n^{(I)}(k) = a_n^{(I)}(0)b_n^{(I)}(k),$$
$$a_n^{(Q)}(k) = a_n^{(Q)}(0)b_n^{(Q)}(k), \qquad k = 1, 2, \ldots, k_u.$$

(4.25)

Furthermore, the timing of all individual users is locked to that of the pilot sequence, so that the multipath delays need only be searched on the pilot sequence.

The optimum demodulator structure for L multipath propagation paths (as assumed in Figure 4.1) is known as a Rake receiver [Price and Green, 1958]. It was first implemented in static form in the late 1950s. This is shown in Figure 4.3 for the kth user. Figure 4.3a consists of the parallel combination of L elements, one of which is shown in Figure 4.3b. Each multipath component demodulator is called a "finger" of the rake. The pilot sequence tracking loop of a particular demodulator is started by the timing delay estimation of a given path, as determined by the pilot's pseudorandom sequence searcher. This is then used to remove the pilot QPSK spreading, giving rise to the quadrature outputs (Figure 4.3):

$$\sqrt{E_c}\,[A_0 + x_n(k)b_n^{(I)}(k)]\alpha_\ell \cos \phi_\ell + \nu_n^{(I)},$$
$$\sqrt{E_c}\,[A_0 + x_n(k)b_n^{(Q)}(k)]\alpha_\ell \sin \phi_\ell + \nu_n^{(Q)}.$$

A_0 is the pilot gain, and $\nu_n^{(I)}$ and $\nu_n^{(Q)}$ are the contributions of all other (uncorrelated) multipath components as well as those of all other users. From this the relative path values $\alpha_\ell \cos \phi_\ell$ and $\alpha_\ell \sin \phi_\ell$ can be estimated by simply averaging over an arbitrary number of chips, N_p. This number should be as large as possible without exceeding the period over which α_ℓ and ϕ_ℓ remain relatively constant.

The optimum (maximum likelihood) demodulator forms the weighted, phase-adjusted, and delay-adjusted sum of the L components. This amounts to taking the inner product of the modulated received I and Q components with the I and Q unmodulated component magnitude estimates $\hat{\alpha}_\ell \cos \hat{\phi}_\ell$ and $\hat{\alpha}_\ell \sin \hat{\phi}_\ell$. The result for the nth chip of the ℓth path, after multiplication by the quadrature-user-specific pseudorandom sequences as shown in Figure 4.3b, is

$$y_{\ell n}(k) = \sqrt{E_c}\,x_n(k)\hat{\alpha}_\ell \alpha_\ell \cos(\phi_\ell - \hat{\phi}_\ell)$$
$$+ \hat{\alpha}_\ell(\nu_n^{(I)} \cos \hat{\phi}_\ell + \nu_n^{(Q)} \sin \hat{\phi}_\ell).$$

(4.26)

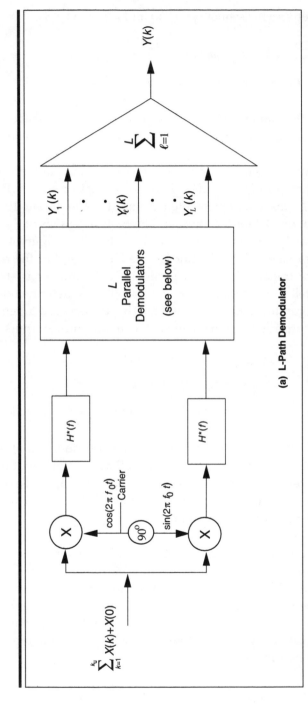

Figure 4.3 Pilot-aided coherent Rake demodulator for multipath propagation. (a) *L*-path demodulator.

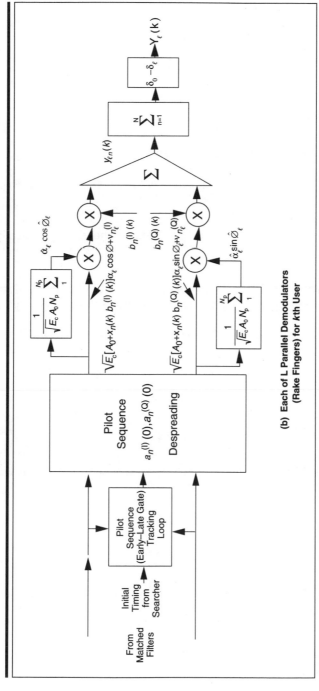

(b) Each of L Parallel Demodulators (Rake Fingers) for _k_th User

Figure 4.3 (b) Each of *L* parallel demodulators (Rake fingers) for *k*th user.

Summing over the N chips over which $x_n(k)$ is constant (± 1), we have

$$
\begin{aligned}
Y_\ell(k) = &\pm N\sqrt{E_c}\, \alpha_\ell \hat{\alpha}_\ell \cos(\phi_\ell - \hat{\phi}_\ell) \\
&+ \hat{\alpha}_\ell [\cos \hat{\phi}_\ell \sum \nu_n^{(I)} + \sin \hat{\phi}_\ell \sum \nu_n^{(Q)}].
\end{aligned}
\tag{4.27}
$$

Thus,

$$
E[Y_\ell(k) \mid x_n(k) = -1] = -N[\sqrt{E_c}\, \alpha_\ell \hat{\alpha}_\ell \cos(\phi_\ell - \hat{\phi}_\ell)], \tag{4.28a}
$$

with a change of sign if $x_n(k) = +1$, and

$$
\mathrm{Var}[Y_\ell(k)] = N\hat{\alpha}_\ell^2 I_0/2, \tag{4.28b}
$$

where I_0 is given in (4.16).

We proceed to obtain Chernoff bounds on the error probability using the methods of Section 4.2.

4.4.1 Chernoff Bounds on Error Probability for Coherent Demodulation with Known Path Parameters

Initially, conditioning on known amplitudes α_ℓ and phases ϕ_ℓ, we obtain the Chernoff bound,

$$
\begin{aligned}
P_E(\boldsymbol{\alpha}, \boldsymbol{\phi}; k) &= \Pr\left[\sum_{\ell=1}^{L} Y_\ell(k) > 0 \mid \boldsymbol{\alpha}, \boldsymbol{\phi}, x(k) = -1 \right] \\
&< \operatorname*{Min}_{\rho>0} E\left(\exp\left[\rho \sum_{\ell=1}^{L} Y_\ell(k) \right] \mid \boldsymbol{\alpha}, \boldsymbol{\phi}, x(k) = -1 \right) \\
&= \operatorname*{Min}_{\rho>0} \exp\left[-\rho N \sqrt{E_c(k)} \sum_{\ell=1}^{L} \alpha_\ell \hat{\alpha}_\ell \cos(\phi_\ell - \hat{\phi}_\ell) + \rho^2 N \sum_{\ell=1}^{L} \hat{\alpha}_\ell^2 I_0/4 \right] \\
&= \exp\left\{ -\frac{NE_c(k) \left[\displaystyle\sum_{\ell=1}^{L} \alpha_\ell \hat{\alpha}_\ell \cos(\phi_\ell - \hat{\phi}_\ell) \right]^2}{\displaystyle\sum_{\ell=1}^{L} \hat{\alpha}_\ell^2 I_0} \right\}.
\end{aligned}
\tag{4.29}
$$

If we neglect the inaccuracy in the amplitude and phase estimates, taking $\hat{\phi}_\ell = \phi_\ell$, $\hat{\alpha}_\ell = \alpha_\ell$, we obtain

$$
\begin{aligned}
P_E(k) &< \exp\left[-\sum_{\ell=1}^{L} \alpha_\ell^2 N E_c(k)/I_0 \right] \\
&= \prod_{\ell=1}^{L} \exp[-\alpha_\ell^2 N E_c(k)/I_0] \quad \text{(perfect estimates)}.
\end{aligned}
\tag{4.30}
$$

Further, it is shown in Appendix 4A that if we do not assume exact phase and amplitude estimates, but rather estimates based on N_p unmodulated chips of a pilot whose chip energy is $A_0^2 E_c$ (see Figure 4.3), the error probability is bounded by

$$P_E(k) < \prod_{\ell=1}^{L} \frac{\exp[-\alpha_\ell^2 N E_c(k)/I_0]}{1 - [N/(A_0^2 N_p)]}. \tag{4.31}$$

Generally, if the paths are known and the total received energy per chip is $E_c(k)$, then we may normalize the relative path gains so that

$$\sum_{\ell=1}^{L} \alpha_\ell^2 = 1.$$

Thus, for fixed amplitude and phase multipath, the performance bound with perfect estimates, (4.30), is the same as for a single-component signal, if energy is taken as the sum of the component energies. The explanation is simple: When the multipath amplitudes and phases are known, the optimal receiver operates as a matched filter to the combination of the transmitter filter and the (multipath) channel.

4.4.2 Rayleigh and Rician Fading Multipath Components

Now we no longer assume constant amplitude. We let the multipath component amplitudes be random variables, mutually independent because we assume that each path's attenuation is unrelated to that of all others. Then the error probability for perfect estimates becomes

$$P_E = E[P_E(\alpha_1, \ldots, \alpha_L)] < E\left[\prod_{\ell=1}^{L} \exp(-\alpha_\ell^2 N E_c/I_0)\right]$$

$$= \prod_{\ell=1}^{L} E[\exp(-\alpha_\ell^2 E_s/I_0)] \triangleq \prod_{\ell=1}^{L} Z_\ell \triangleq Z. \tag{4.32}$$

Here, $E_s \triangleq N E_c$ is the N-chip *symbol* energy, and the expectations are with respect to the random variables α_ℓ. We drop the user index k for convenience. We also assume perfect estimates, although by scaling all Z_ℓ by the denominator of (4.31), we may also obtain a bound for imperfect estimates.

If each component is the combination of many reflections arriving at nearly the same delay but with random phases, we can take the α_ℓ variable to be Rayleigh-distributed. Then the probability density function of α_ℓ is

$$p(\alpha) = \frac{2\alpha \, e^{-\alpha^2/\sigma_\ell^2}}{\sigma_\ell^2}, \qquad \alpha > 0. \tag{4.33a}$$

Or, letting $\beta_\ell = \alpha_\ell^2$, we obtain the chi-squared density,

$$p(\beta) = \frac{e^{-\beta/\sigma_\ell^2}}{\sigma_\ell^2}, \qquad \beta > 0, \tag{4.33b}$$

where $\sigma_\ell^2 = E[\beta_\ell] = E[\alpha_\ell^2]$.

Thus, for Rayleigh-distributed attenuations,

$$Z_\ell = E[e^{-\beta_\ell E_s/I_0}] = \int_0^\infty \frac{1}{\sigma_\ell^2} \exp\left[\frac{-\beta}{\sigma_\ell^2}\left(\frac{E_s}{I_0}\sigma_\ell^2 + 1\right) \right] d\beta$$

$$= \frac{1}{1 + \sigma_\ell^2 E_s/I_0}. \tag{4.34}$$

Letting

$$\overline{E}_{s_\ell} = \overline{\beta_\ell} E_s = \sigma_\ell^2 E_s,$$

this can be written as

$$Z_\ell = \frac{1}{1 + \overline{E}_{s_\ell}/I_0} \qquad \text{(Rayleigh fading component).} \tag{4.35}$$

If the component is the combination of a specular component and a Rayleigh component, the probability density function of α_ℓ becomes Rician. Its square β_ℓ becomes noncentral chi-squared,

$$p(\beta_\ell) = \frac{e^{-(\beta_\ell + \gamma_\ell)/\sigma_\ell^2}}{\sigma_\ell^2} \mathcal{I}_0(2\sqrt{\gamma_\ell \beta_\ell}/\sigma_\ell^2). \tag{4.36}$$

Then

$$Z_\ell = E[e^{-\beta_\ell E_s/I_0}]$$

$$= \int_{-\infty}^\infty \exp\left[-\frac{\gamma_\ell + \beta_\ell(1 + \sigma_\ell^2 E_s/I_0)}{\sigma_\ell^2} \right] \mathcal{I}_0\left(2\frac{\sqrt{\gamma_\ell \beta_\ell}}{\sigma_\ell^2} \right) d\beta_\ell/\sigma_\ell^2$$

$$= \frac{1}{1 + \sigma_\ell^2 E_s/I_0} \exp\left(-\frac{\gamma_\ell E_s/I_0}{1 + \sigma_\ell^2 E_s/I_0} \right) \qquad \text{(Rician fading component).} \tag{4.37}$$

Note that this reduces to the Rayleigh fading result (4.34) when $\gamma_\ell = 0$, and to the known amplitude and phase result when $\sigma_\ell^2 = 0$.

Finally, suppose that the L multipath components are *all Rayleigh of equal average strength*, so that

$$\sigma_\ell^2 = \sigma^2 \qquad \text{for all } \ell,$$

and therefore, for each component,

$$\overline{E}_{s_\ell} = \sigma_\ell^2 E_s = \sigma^2 E_s = \overline{E}_s.$$

Then, letting the random variable

$$x = \sum_{\ell=1}^{L} \alpha_\ell^2 E_s = \sum_{\ell=1}^{L} \beta_\ell E_s,$$

it is easily shown (see Appendix 3A.5) that since the individual fading variables are all independent,

$$p(x) = \frac{x^{L-1} e^{-x/\overline{E}_s}}{(L-1)!(\overline{E}_s)^L}, \tag{4.38}$$

which is Lth-order chi-squared. It follows from (4.34) that, in this case,

$$\overline{P}_E < Z \triangleq \prod_{\ell=1}^{L} Z_\ell = \left[\frac{1}{1 + (\sigma^2 E_s / I_0)} \right]^L = \left[\frac{1}{1 + (\overline{E}_s / I_0)} \right]^L. \tag{4.39}$$

We may rewrite (4.39) as

$$\overline{P}_E < \exp[-\ln(1/Z)],$$

where

$$\ln(1/Z) = L \ln[1 + (\overline{E}_s / I_0)]. \tag{4.40}$$

From (4.40) we obtain Figure 4.4: a plot of the ratio[6] of the total average symbol energy-to-interference density $L\overline{E}_s / I_0$ over that required for an unfaded Gaussian channel to achieve a given exponent value, $\ln(1/Z)$. Note from (4.9) or (4.32) that for the latter channel, $\ln(1/Z) = E_s / I_0$. Thus, this is the average excess energy (in decibels) required by this degraded channel to achieve the same performance as for an unfaded signal in additive Gaussian noise.

---

[6] All quantities are in decibels and hence are logarithmic functions. Thus, this ratio is actually the excess energy, in decibels, required for the faded channel over that required for the unfaded AWGN channel.

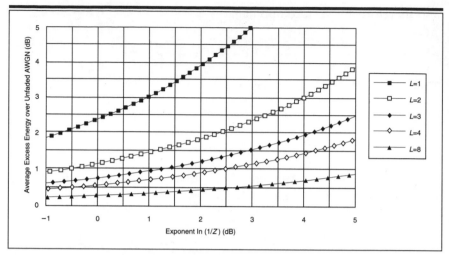

Figure 4.4 Required excess energy (dB) for *L* equal-strength multipath fading relative to unfaded coherent AWGN.

Note that as $L \to \infty$ (so that each component's average energy $\overline{E}_s \to 0$, but $L\overline{E}_s$ is finite), the bound approaches

$$\overline{P}_E < e^{-L\overline{E}_s/I_0},$$

so that the excess energy approaches zero. This shows that with an asymptotically large number of independent Rayleigh components, performance approaches that of unfaded propagation. This is an extreme and unrealistic example of the beneficial effect of independent diversity components. We shall return to these results in Chapter 5 when we consider interleaving, with delay, to produce more independent components.

4.5 Noncoherent Reception

Transmission of a pilot is very valuable for initial acquisition and time tracking. It is also valuable for obtaining good amplitude and phase estimates, making possible quasi-optimum coherent reception and weighted combining of multipath components. Unfortunately, it is a luxury that may not always be affordable, particularly on the many-to-one reverse (up-) link from each of the multiple access users to the base station. For this purpose, inserting in each individual user's signal a pilot whose power is greater than the data-modulated portion of the signal reduces efficiency to less than 50%. On the other hand, without phase and amplitude estimation, noncoherent or differentially coherent reception is required. Timing of new paths must still be acquired and tracked—a more difficult task without a pilot. We address this in the next section. In the present section we assume that this timing is available, but phase and amplitude estimates are not.

4.5.1 Quasi-optimum Noncoherent Multipath Reception for *M*-ary Orthogonal Modulation

Suppose that L independent multipath components are being tracked at a given time. Suppose also that neither phase nor amplitude estimates are available, but their average relative powers are assumed to be equal. These components may actually be received by a combination of two or more antennas, since spatial diversity antennas are commonly used at the base station. As for the coherent case, we assume that a separate demodulator is provided for each path, but now their outputs are noncoherently combined.

For noncoherent demodulation we cannot employ antipodal signals, $x_n(k)$ equal to $+1$ or -1, remaining constant for the duration of the N chips constituting a symbol, because the quadratic operations inherent in noncoherent demodulation destroy signs. The best we can accomplish on the basis of a single symbol is to make the "0" and "1" waveforms orthogonal. Keeping the same modulator–demodulator structure as in Figure 2.4, we can achieve this by mapping a "0" into a binary signal consisting of N chips with positive sign and mapping a "1" into a binary signal consisting of $N/2$ positive chips and $N/2$ negative chips, where N is even.

That is,

$$x_n(k) = +1, \qquad n = 1, 2, \ldots, N, \qquad \text{if "0" sent,}$$

and

$$x_n(k) = \begin{cases} +1, & n = 1, 2, \ldots, N/2 \\ -1, & n = N/2 + 1, \ldots, N \end{cases} \quad \text{if "1" sent,}$$

so that the two signals are orthogonal over the N chips.

There are two alternatives to simple binary orthogonal modulation. The first is to use *differential binary modulation (DPSK)*, where

$$x_n(k) = \begin{cases} x_{n-1}(k) \text{ if present dada symbol is "0"} \\ -x_{n-1}(k) \text{ if present data symbol is "1"} \end{cases}$$

It can easily be shown [Wozencraft and Jacobs, 1965; Viterbi, 1966] that the performance is the same as for binary orthogonal modulation, but with *symbol energy doubled*. While this makes the differential approach attractive, even better noncoherent performance can be obtained with M-ary orthogonal modulation, defined as follows.

Suppose we collect $J = \log_2 M$ data symbols to be transmitted, where J is an integer and hence M is a power of 2. Then, provided $N = M$ or a multiple thereof, we proceed to transmit one of M orthogonal binary sequences, known as Hadamard–Walsh functions. The sequences are constructed according to the algorithm described by the example of Figure 4.5 and the general formula shown at the bottom of the array. The algorithm may be implemented by the device whose block diagram is shown in Figure 4.6. It is easily verified that any two vectors resulting from this Hadamard–Walsh mapping have inner products that are exactly zero, and hence all M waveforms are orthogonal to one another.

For this encoder, we denote the input binary data symbols by $\xi_1, \ldots,$ $\xi_2, \ldots, \xi_j$, and the output binary symbols to be transmitted as $x_1, \ldots,$ x_M. N must be an integer multiple ℓ of M. Then each of the M binary symbols of the Hadamard functions is of duration $T = \ell T_c$ and encompasses ℓ successive chips of the pseudorandom sequence signal.

Note that although this orthogonal modulation is much more elaborate[7] than the binary PSK modulation used with coherent demodulation (and which can be shown to be optimum for that case), exactly the same basic waveform and spreading are employed (Figure 2.4a). This is an indication of the flexibility and generality of direct sequence spread spectrum techniques.

The optimum noncoherent demodulator for Hadamard–Walsh orthogonal modulation is a bank of M orthogonal noncoherent correlators [Proakis, 1989], one of which is shown in Figure 4.7b. Each of the M non-

---

[7] It represents, in fact, a block error-correcting code superimposed on the basic modulation. We defer until the next chapter, however, a more extensive treatment of forward error correction.

Symbol	Sequence	Hadamard Mapping
$\xi_1 = 0\,0\,0$		+ + + + + + + +
$\xi_2 = 0\,0\,1$		+ - + - + - + -
$\xi_3 = 0\,1\,0$		+ + - - + + - -
$\xi_4 = 0\,1\,1$		+ - - + + - - +
$\xi_5 = 1\,0\,0$		+ + + + - - - -
$\xi_6 = 1\,0\,1$		+ - + - - + - +
$\xi_7 = 1\,1\,0$		+ + - - - - + +
$\xi_8 = 1\,1\,1$		+ - - + - + + -

Note: To generalize for all $M = 2^J$

$$[H_M] = \begin{bmatrix} H_{M/2} & H_{M/2} \\ H_{M/2} & -H_{M/2} \end{bmatrix}$$

Figure 4.5 Hadamard–Walsh orthogonal sequences for $M = 8$, $J = 3$.

coherent correlators consists of the generic demodulator shown in Figure 2.4b. The "signal processor" block consists of squarers for each of the two quadrature components, followed by an adder. For a single path ($L = 1$), this demodulator makes its decision by selecting the largest of the M output magnitudes as corresponding to the sequence most likely to have been transmitted. It is well known [Helstrom, 1968; Proakis, 1989] that for

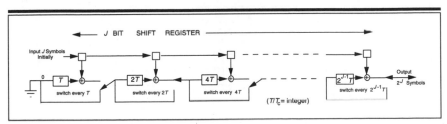

Figure 4.6 Hadamard–Walsh orthogonal sequence generator.

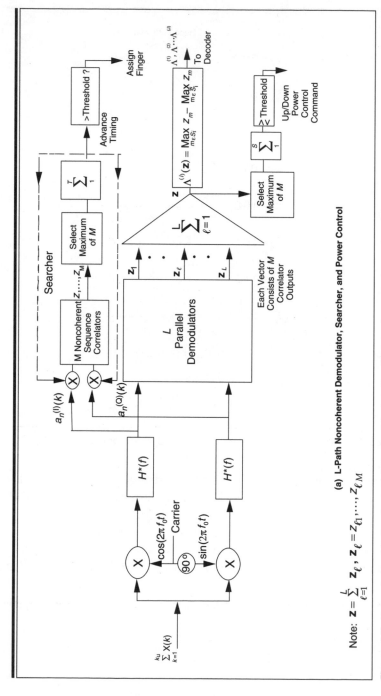

Figure 4.7 Noncoherent Rake demodulator for *M*-ary orthogonal waveforms in *L*-component multipath. (a) *L*-path noncoherent demodulator, searcher, and power control.

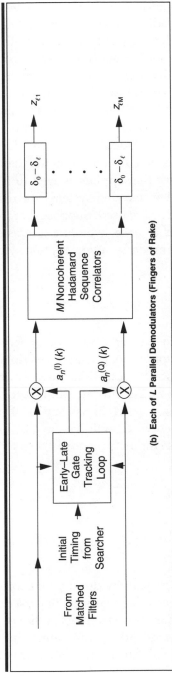

Figure 4.7 (*Cont.*) (b) Each of *L* parallel demodulators (fingers of Rake).

$L = 1$, this is optimum in additive Gaussian noise both for fixed unfaded signals and for Rayleigh fading signals. In the Rayleigh fading case, if the L multipath components have equal average strength, the optimum decision requires adding the individual noncoherent correlator outputs for each of L independent paths before a decision is made, as shown in Figure 4.7a. In the unfaded case, a nonlinear combination is optimum [see (3A.18) in Appendix 3A.4], but at low E_s/I_0 per path, the linear combining only slightly degrades performance. Thus, we shall also assume this slightly suboptimal receiver for the unfaded L-path case.

Our treatment in this section deviates from the standard one to provide a more general framework that applies as well to coded systems to be considered in Chapter 5. For uncoded systems, the result is the same as if the conventional approach were used. Suppose that rather than making a decision on all J symbols at once, we were interested only in the *first* symbol. Then the (optimum) decision for this first symbol is based on the log likelihood ratio

$$\ln[p_0^{(1)}(\mathbf{z})/p_1^{(1)}(\mathbf{z})].$$

$\mathbf{z} = (z_1, z_2, \ldots , z_M)$ is the vector whose components are the M noncoherent (quadratic) correlator outputs. The superscript indicates the first of J symbols ξ_1, and the subscripts of the likelihood functions indicate that a "0" or a "1" was sent for ξ_1. As indicated in Figure 4.5, the first $M/2 = 2^{J-1}$ sequences $\boldsymbol{\xi}_m$ ($m = 1, \ldots , M/2$) will have a "0" in the first position, while the last $M/2$ sequences $\boldsymbol{\xi}_m$ will have a "1" in the first position. Hence, given equiprobable input sequences,

$$p_0^{(1)}(\mathbf{z}) = \frac{2}{M} \sum_{m=1}^{M/2} p(\mathbf{z}/\boldsymbol{\xi}_m), \tag{4.41}$$

$$p_1^{(1)}(\mathbf{z}) = \frac{2}{M} \sum_{m=1+M/2}^{M} p(\mathbf{z}/\boldsymbol{\xi}_m). \tag{4.42}$$

Here, $p(\mathbf{z}/\boldsymbol{\xi}_m)$ is the joint probability density function of the M correlator outputs, given that data vector $\boldsymbol{\xi}_m$ was sent. Superscript (1) indicates that the likelihood function is for the first symbol of the sequence $\boldsymbol{\xi}$. Because of the orthogonality of the sequences representing the ξ's, we may write these joint probability density functions of the correlator outputs as

$$p(\mathbf{z}/\boldsymbol{\xi}_m) = p_C(z_m) \prod_{m' \neq m} p_I(z_{m'}). \tag{4.43}$$

$p_C(z)$ is the probability density function of the correlator output corresponding to the (correct) signal sent, while $p_I(z)$ pertains to all the $M - 1$ (incorrect) others. Thus, $p_C(z)$ refers to the presence of a (correct) signal in the given correlator, while $p_I(z)$ refers to its absence. Hence, these are the same as $p_1(z)$ and $p_0(z)$ of the last chapter (Appendix 3A). Here, L refers to the *number of independent paths,* rather than the number of independent observations.

Then, for L *Rayleigh fading* components of multipath with additive white Gaussian noise, the sum of the L components will have density functions (see Appendix 3A.5 with $\bar{\mu} = J\bar{E}_s/I_0$)

$$p_I(z) = \frac{z^{L-1}}{(L-1)!}\exp(-z), \tag{4.44}$$

$$p_C(z) = \frac{z^{L-1}}{(L-1)!}\frac{\exp[-z/(1 + J\bar{E}_s/I_0)]}{[1 + J\bar{E}_s/I_0]^L}. \tag{4.45}$$

$\bar{E}_s = N\bar{E}_c$ is the average symbol energy per path, and all the correlator outputs are normalized so that all the incorrect ones have unit variance. For L equal *unfaded* paths,[8] (see Appendix 3A.4 with $\mu = JE_s/I_0$),

$$p_C(z) = \left(\frac{z}{LJE_s/I_0}\right)^{(L-1)/2}\exp(-z - LJE_s/I_0)\,\mathcal{I}_{L-1}(2\sqrt{z\,LJE_s/I_0}). \tag{4.46}$$

E_s is the symbol energy per path, and, of course, $p_I(z)$ is again given by (4.44).

Now from (4.41), (4.42), and (4.43), it follows upon dividing both numerator and denominator by $\Pi_{m'}\, p_I(z_{m'})$ that the log likelihood ratio for the first symbol is

$$\ln\left[\frac{p_0^{(1)}(\mathbf{z})}{p_1^{(1)}(\mathbf{z})}\right] = \ln\left[\frac{\displaystyle\sum_{m=1}^{M/2} p_C(z_m)/p_I(z_m)}{\displaystyle\sum_{m=1+M/2}^{M} p_C(z_m)/p_I(z_m)}\right]. \tag{4.47}$$

We proceed to consider only L-path Rayleigh fading, since this represents a worst case. Also, as noted, only in the case of Rayleigh fading with

---

[8] As previously noted, the summation of energies is suboptimal in this case, but the optimal (nonlinear) combining requires weighing each component with its energy-to-interference level, which is generally unknown [see (3A.18) in Appendix 3A.4]. For low E_s/I_0 per path, the effect of suboptimal linear combining is negligible.

independent equal-average energy paths does the summing of the path energies represent the optimum combining function. Thus, from (4.44), (4.45), and (4.47), we obtain

$$
\ln\left[\frac{p_0^{(1)}(\mathbf{z})}{p_1^{(1)}(\mathbf{z})}\right] = \ln\frac{\displaystyle\sum_{m=1}^{M/2}\exp\left[\frac{z_m\,\overline{J E_s}/I_0}{1+\overline{J E_s}/I_0}\right]}{\displaystyle\sum_{m=1+M/2}^{M}\exp\left[\frac{z_m\,\overline{J E_s}/I_0}{1+\overline{J E_s}/I_0}\right]}.
\qquad (4.48)
$$

In this case and for virtually any channel statistics, the log likelihood ratio is too complex to compute for large M, and it requires knowledge of $\overline{E_s}/I_0$. On the other hand, since the exponential is a rapidly growing function of its argument, the sums in both the numerator and denominator of the likelihood ratio are dominated by the respective maxima of the $M/2$ terms. Thus, for the L-path Rayleigh fading case, letting $K_f = (\overline{J E_s}/I_0)/(1 + \overline{J E_s}/I_0)$,

$$
\ln\left[\frac{p_0^{(1)}(\mathbf{z})}{p_1^{(1)}(\mathbf{z})}\right] \approx \ln\left[\frac{\displaystyle\operatorname*{Max}_{m=1}^{M/2}\exp(K_f z_m)}{\displaystyle\operatorname*{Max}_{m=1+M/2}^{M}\exp(K_f z_m)}\right]
$$

$$
= K_f\left[\operatorname*{Max}_{m=1}^{M/2} z_m - \operatorname*{Max}_{m=1+M/2}^{M} z_m\right].
$$

Thus, we use the nonparametric decision metric,[9]

$$
\Lambda^{(1)}(\mathbf{z}) = \operatorname*{Max}_{m=1}^{M/2} z_m - \operatorname*{Max}_{m=1+M/2}^{M} z_m.
$$

We proceed in the same way for the other symbols of the sequence

$$
\boldsymbol{\xi} = \xi_1, \xi_2, \ldots, \xi_J.
$$

In general, let

$$
S_i = \{\text{all } m: i\text{th component of } \xi_m \text{ is ``0''}\}
\qquad (4.49)
$$

---

[9] Even with knowledge of $\overline{E_s}/I_0$, or E_s/I_0, and utilizing the optimum processor defined by (4.47), the performance improvement is only on the order of a few tenths of a decibel, determined by simulation. Also, while this is not the only way to generate a soft decision metric for M-ary noncoherent reception, it outperforms earlier metric choices [Viterbi *et al.*, 1993].

and consequently,

$$\overline{S}_i = \{\text{all } m: i\text{th component of } \boldsymbol{\xi}_m \text{ is "1"}\}. \tag{4.50}$$

For example, for the second data symbol, $S_2 = \{1, 2, \ldots, M/4, 1 + M/2, \ldots, 3M/4\}$, while for the last data symbol, $S_J = \{\text{all odd integers} < M\}$. Then proceeding as before,

$$\Lambda^{(i)}(\mathbf{z}) = \underset{m \in S_i}{\text{Max }} z_m - \underset{m \in \overline{S}_i}{\text{Max }} z_m. \tag{4.51}$$

The decision is then made for a "0" if $\Lambda^{(i)}(\mathbf{z}) \geq 0$ and for a "1" otherwise.

Before we analyze the performance of this approximation to the optimum symbol decision metric, we note that the conventional approach chooses the J symbols at once as those pertaining to the single maximum correlator output from the set of $M = 2^J$ outputs $z_1, \ldots, z_M$. This produces the most likely symbol sequence, $\boldsymbol{\xi}_1, \ldots, \boldsymbol{\xi}_J$, and no approximations are required. It is obvious that the conventional approach yields exactly the same decisions as (4.51). But this alternate symbol-by-symbol decision approach will provide a basis for achieving better performance with error-correcting codes, by using "soft decisions," as will be shown in Chapter 5.

4.5.2 Performance Bounds

On a symbol-by-symbol basis, the Chernoff bound on error probability is obtained as for the coherent case: by bounding the probability that the log likelihood function (or its approximation that will be used as the metric) is negative if a "0" was sent or positive if a "1" was sent. Thus, from (4.51) we have that, for equiprobable hypotheses, for all $i, i = 1, \ldots, J$,

$$
\begin{aligned}
P_E^{(i)} &= \Pr[\Lambda^{(i)}(\mathbf{z}) > 0 \mid \text{"1"}] = \Pr[\Lambda^{(i)}(\mathbf{z}) < 0 \mid \text{"0"}] \\
&= \Pr\left[\underset{m \in S_i}{\text{Max }} z_m - \underset{m \in \overline{S}_i}{\text{Max }} z_m > 0 \mid \text{"1"} \right] \\
&= \Pr\left[\underset{m \in \overline{S}_i}{\text{Max }} z_m - \underset{m \in S_i}{\text{Max }} z_m > 0 \mid \text{"0"} \right]. \tag{4.52}
\end{aligned}
$$

In either case, the distribution of the first "Max" in each bracket is the distribution of the maximum of $M/2$ incorrect correlator outputs. The distribution of the second "Max" in each bracket is the distribution of the maximum of $(M/2) - 1$ incorrect correlator outputs and the correct correlator output. Also note that the two maxima are independent, because they are maxima of disjoint sets of mutually independent variables. Thus, we define

$$
F_I(x) \triangleq \Pr\left[\underset{m \in S_i}{\text{Max } z_m} < x \mid \text{``1''}\right] = \Pr\left[\underset{m \in \bar{S}_i}{\text{Max } z_m} < x \mid \text{``0''}\right]
$$

$$
= \left[\int_0^x p_I(z)\, dz\right]^{M/2} = P_I^{M/2}(x),
\tag{4.53}
$$

$$
F_C(y) \triangleq \Pr\left[\underset{m \in \bar{S}_i}{\text{Max } z_m} < y \mid \text{``1''}\right] = \Pr\left[\underset{m \in S_i}{\text{Max } z_m} < y \mid \text{``0''}\right]
$$

$$
= \int_0^y p_C(z)\, dz \left[\int_0^y p_I(z)\, dz\right]^{M/2-1} = P_C(y)P_I(y)^{M/2-1},
\tag{4.54}
$$

where $p_C(z)$ and $p_I(z)$ are the probability density functions of the correct and of the incorrect correlator outputs, respectively. (Examples of these are given in (4.44) through (4.46).) $P_C(z)$ and $P_I(z)$ are their integrals, the corresponding distribution functions.

Now, applying the Chernoff bound to (4.52), using (4.53) and (4.54), we have that for all i,

$$
P_E < E \exp\left\{\rho\left[\underset{m \in S_i}{\text{Max } z_m} - \underset{m \in \bar{S}_i}{\text{Max } z_m} \mid \text{``1''}\right]\right\}
$$

$$
= E \exp\left\{\rho\left[\underset{m \in \bar{S}_i}{\text{Max } z_m} - \underset{m \in S_i}{\text{Max } z_m} \mid \text{``0''}\right]\right\}
$$

$$
= E \exp[\rho(x - y)]
$$

$$
= \int_0^\infty \exp(\rho x)\, dF_I(x) \int_0^\infty \exp(-\rho y)\, dF_C(y)
\tag{4.55}
$$

$$
= \int_0^\infty \exp(\rho x) \left[\frac{M}{2} P_I^{M/2-1}(x) p_I(x)\right] dx
$$

$$
\times \int_0^\infty \exp(-\rho y)[(M/2 - 1)P_C(y)P_I(y)^{M/2-2}p_I(y)
$$

$$
+ p_C(y)P_I(y)^{M/2-1}]\, dy,
$$

where $\rho > 0$ and

$$
P_I(x) = \int_0^x p_I(z)\, dz \quad \text{and} \quad P_C(x) = \int_0^x p_C(z)\, dz.
\tag{4.56}
$$

In some cases, the integrals in (4.55) can be expressed as finite sums of M terms. In most cases, however, numerical integration is either required or simpler to compute than the sums. We express the minimized bound as in Section 4.4,

$$P_E < \exp[-\ln(1/Z)], \qquad (4.57)$$

where Z is given by (4.55) minimized over $\rho > 0$. Using (4.44) though (4.46) in (4.55), we evaluate Z as a function of LJE_s/I_0 (or $LJ\overline{E}_s/I_0$ in the faded case).

We then proceed as for the coherent case (Figure 4.4). For each value in Figure 4.8 of LJE_s/I_0 and the corresponding value of Z, we plot $(LJE_s/I_0)/\ln(1/Z)$ in decibels, which is the excess energy required relative to an unfaded BPSK signal coherently demodulated in AWGN (for which $\ln(1/Z) = LJE_s/I_0$). We do this for M = 64 orthogonal sequences ($J = 6$) and for $L = 1$ and $L = 2$ components of multipath. A number of interesting conclusions can be drawn from this figure. First, two fixed paths, in AWGN only, perform worse than one fixed path with the same total energy because of the *noncoherent combining loss*. However, two paths that are Rayleigh-faded perform better (for all but very low energy-to-noise) than one path with the same total energy. This is because *the diversity gain is greater than the noncoherent combining loss*. AWGN with noncoherent demodulation of M-ary orthogonal modulation performs better than coherently demodulated BPSK at higher energy-to-interference levels (the decibel value of the ratio goes negative) because the M-ary system com-

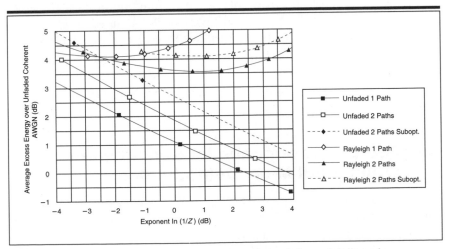

Figure 4.8 Required excess E_b/I_0 (dB) for noncoherent demodulation of orthogonal signals ($M = 64$).

bines block (orthogonal) coding with the modulation. As we shall see in Chapter 5, however, for equivalent coding complexities, coherently demodulated BPSK always performs better than noncoherent demodulation of any form of modulation. This is related to the fact that, with coding, smaller values of $\ln(1/Z)$ are required, which is particularly evident with lower rate codes.

One more alternative needs to be considered. In the noncoherent case in each correlator, we have summed the energies from the L paths before forming the decision variable $\Lambda^{(i)}(\mathbf{z})$ according to (4.51). That is, each of the variables z_m consisted of the sum of L noncoherent correlator outputs, one for each multipath component (Figure 4.7). Alternatively, we could form the decision variable $\Lambda^{(i)}(\mathbf{z})$ for each path individually and then sum the L variables, making a decision based on this sum of the L individual path decision variables. In this case the error probability bound in place of (4.55) becomes

$$P_E < E \exp\left\{\rho \sum_{\ell=1}^{L}\left[\underset{m \in S_i}{\text{Max }} z_m^{(\ell)} - \underset{m \in \bar{S}_i}{\text{Max }} z_m^{(\ell)} \mid \text{"1"}\right]\right\}, \qquad \rho > 0,$$

where $z_m^{(\ell)}$ is the mth correlator output for the ℓth path. Then, since all paths are mutually independent, for equal path energies,

$$P_E < Z_1^L = \exp[-L\ln(1/Z_1)] \triangleq Z \qquad \text{(suboptimal combining)}, \quad (4.58)$$

where Z_1 is the same as given in (4.55), but for the case of $L = 1$ path. In this case, the ratio of LJE_s/I_0 to $\ln(1/Z)$, as a function of $\ln(1/Z)$, is obtained by taking the $L = 1$ path results previously obtained and multiplying both coordinates by L (actually adding $10\log_{10}L$ to the decibel values). We denote this approach and the corresponding curves (shown dotted in Figure 4.8) as suboptimal combining. The degradation due to this suboptimal combining for $L = 2$ is about 0.8 dB for AWGN and about 0.5 dB for Rayleigh fading multipath.

4.6 Search Performance for Noncoherent Orthogonal *M*-ary Demodulators

In the noncoherent demodulator of Figure 4.7 just considered, each of the multipath components must be searched without the benefit of the unmodulated pilot sequence. Thus, the search must be performed on modulated signals as shown in the upper part of Figure 4.7a. The transmitted

Hadamard sequence is most likely to correspond to the correlator with the largest output, so the searcher bases its timing decision on the sum of T successive maxima. If the timing is correct, within less than one chip duration, the probability distribution function of each maximum of the M correlators is

$$F_S(z) = P_C(z)P_I(z)^{M-1}. \qquad (4.59)$$

$P_C(z)$ and $P_I(z)$ are the distribution functions of the correct and of the incorrect correlator outputs, respectively (with $L = 1$ component because the search is performed on each multipath component individually). If the timing error is more than one chip duration, based on the delay-and-add property (2.10), all correlators will observe essentially random sequence inputs only. Hence, the distribution function of the maximum in this case is

$$F_0(z) = P_I^M(z). \qquad (4.60)$$

Based on observation of the sum of T independent successive maxima, the optimum decision, according to both Bayes and Neyman–Pearson criteria (Appendix 3A), is based on the log-likelihood ratio,

$$
\begin{aligned}
\Lambda(\mathbf{z}) &= \sum_{t=1}^{T} \ln \frac{dF_S(z_t)/dz_t}{dF_0(z_t)/dz_t} \\
&= \sum_{t=1}^{T} \ln \frac{(M-1)P_C(z_t)P_I(z_t)^{M-2}p_I(z_t) + p_C(z_t)P_I(z_t)^{M-1}}{M\, P_I^{M-1}(z_t)p_I(z_t)} \qquad (4.61) \\
&= \sum_{t=1}^{T} \ln \left[\left(\frac{M-1}{M}\right)\frac{P_C(z_t)}{P_I(z_t)} + \left(\frac{1}{M}\right)\frac{p_C(z_t)}{p_I(z_t)} \right].
\end{aligned}
$$

Here, $\mathbf{z}$ is the vector consisting of the T successive maximum values, normalized by automatic gain control (AGC) so that incorrect correlator outputs have unit variance. Each term of the log-likelihood ratio metric is plotted in Figures 4.9a and 4.9b as a function of the maximum correlator output z. These represent performance for unfaded and for Rayleigh fading signals, respectively, for $M = 64$ and orthogonal-signal[10] energy-to-interference, $E/I_0 = 6$ dB and 9 dB, including any degradation due to timing error of less one chip duration.

---

[10] E is the energy in the signal representing J successive symbols, according to the mapping described in Section 4.5.1.

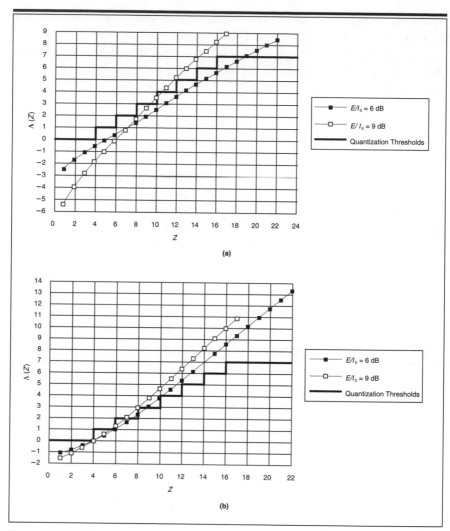

Figure 4.9 (a) Unfaded energy signal log-likelihood function ($M = 64$). (b) Rayleigh fading signal log-likelihood function ($M = 64$).

The false alarm and detection probabilities are the probabilities that $\Lambda(\mathbf{z})$ of (4.61) exceeds a given threshold ψ under the hypotheses, respectively, that signal is absent (i.e., more than one chip duration in error) and that it is present (no timing error).

Thus,

$$P_F = \Pr(\Lambda(\mathbf{z}) > \psi \mid \text{signal absent}),$$
$$P_D = \Pr(\Lambda(\mathbf{z}) > \psi \mid \text{signal present}). \tag{4.62}$$

We could derive a Chernoff bound for these expressions. However, it is more useful to obtain exact expressions for the case where the likelihood ratio is quantized, in accordance with the requirements of a practical digital signal processor. Figures 4.9a and b show that the uniform step quantizer with eight levels is a reasonable approximation for the nearly linear region of the log-likelihood ratio metric for both the faded and unfaded cases. Note also that the quantizer was chosen to be nonparametric, while the likelihood ratio requires knowledge of E/I_0.

With the maxima thus quantized, we obtain the metric

$$\hat{\Lambda}(\mathbf{z}) = \sum_{t=1}^{T} \Lambda_Q(z_t). \tag{4.63}$$

$\Lambda_Q(z_t)$ is the quantized value of the log-likelihood value of each symbol, which is an integer defined as (see Figure 4.9a and b)

$$\Lambda_Q(z) = \begin{cases} 0 & \text{if } z < 4, \\ \lceil (z-4)/2 \rceil & \text{if } 4 \le z \le 16, \\ 7 & \text{if } z > 16. \end{cases} \tag{4.64}$$

$\lceil x \rceil$ indicates the least integer not smaller than x. This, of course, corresponds to the thresholds $\theta_k = 4 + 2(k-1)$, $k = 1, 2, \ldots, 7$. As a result, $\Lambda_Q(z)$ takes on integer values 0 to 7 with probabilities, in the absence and presence of signal, respectively,

$$Q_k^{(0)} = \int_{\theta_k}^{\theta_{k+1}} dF_0(z) = P_1^M(z)\Big|_{\theta_k}^{\theta_{k+1}}, \qquad k = 0, 1, 2, \ldots, 7,$$

$$\tag{4.65}$$

$$Q_k^{(S)} = \int_{\theta_k}^{\theta_{k+1}} dF_S(z) = P_C(z)P_1^{M-1}(z)\Big|_{\theta_k}^{\theta_{k+1}}, \qquad k = 0, 1, 2, \ldots, 7,$$

with $\theta_0 = 0$, $\theta_8 = \infty$.

Then, with metrics quantized in this manner,

$$P_F = \Pr\left(\sum_{t=1}^{T} \Lambda_Q(z_t) > \psi \,|\, \text{signal absent}\right)$$

and

$$P_D = \Pr\left(\sum_{t=1}^{T} \Lambda_Q(z_t) > \psi \,|\, \text{signal present}\right). \tag{4.66}$$

If we assume that successive symbols are independent (always true in unfaded environments), the discrete distribution of the sum of quantized values can be obtained from the (discrete) convolution of the individual quantized values. Better, they can be obtained from the moment-generating function of the sum, which is the product of the generating functions of the individual terms. Thus, we define

$$\pi_0(w) = \sum_{k=0}^{7} Q_k^{(0)} w^k,$$
$$\pi_S(w) = \sum_{k=0}^{7} Q_k^{(S)} w^k.$$

(4.67)

Raised to the Tth power, this is

$$[\pi(w)]^T = Q_0^T + TQ_0^{T-1}Q_1 w + \cdots + Q_7^T w^{7T}, \qquad (4.68)$$

with the appropriate subscripts and superscripts (0 or S). The false alarm and detection probabilities are just the sum of coefficients of the polynomials of (4.68) whose powers exceed the threshold ψ.

Thus,

$$P_F = \{[\pi_0(w)]^T\}_{\psi^+}$$
$$P_D = \{[\pi_S(w)]^T\}_{\psi^+}$$

(4.69)

where $\{ \}_{\psi^+}$ is the sum of the coefficients of all terms whose (integer) powers (of w) exceed ψ. Figure 4.10 shows P_D as a function of P_F (receiver operating characteristic) for various values of $E/I_0 = JNE_c/I_0$ with $T = 2$, 3, and 6. For comparison, the case for unmodulated pilot detection, with $T = 1$, is also shown.

We note finally that the search range, or window, on multipath components should be relatively small once the first component is located. We also note that the first component is generally detected from a (possibly unmodulated) access request transmission (see Chapter 6). This implies that the number of hypotheses to be tested is small, providing for rapid return to the correct hypothesis if it is incorrectly dismissed. Nevertheless, for users in motion, additional multipath components appear and disappear often enough that parallel searchers may be required to quickly find new multipath components and assign demodulator fingers when they appear.

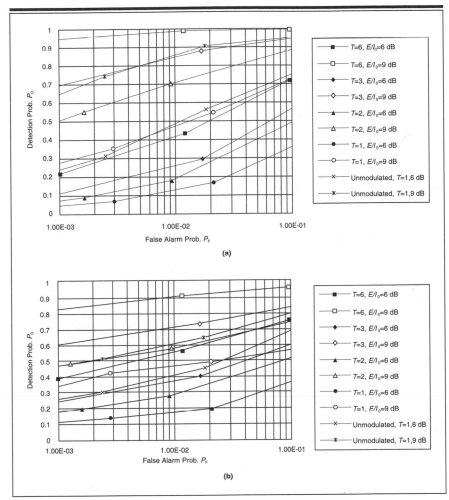

Figure 4.10 (a) False alarm and detection probabilities for search of unfaded components. (b) False alarm and detection probabilities for search of Rayleigh fading components.

4.7 Power Measurement and Control for Noncoherent Orthogonal *M*-ary Demodulators

Power control[11] is a valuable asset in any two-way communication system. It is particularly important in a multiple access terrestrial system where users' propagation loss can vary over many tens of decibels. As

---

[11] The system and network aspects of power control will be treated in Chapter 6. This section is devoted to its implementation with noncoherent reception.

was first discussed in Chapter 1, the power at the cellular base station received from each user over the reverse link must be made nearly equal to that of all others in order to maximize the total user capacity of the system. Very large disparities are caused mostly by widely differing distances from the base station and, to a lesser extent, by shadowing effects of buildings and other objects. Such disparities can be adjusted individually by each mobile subscriber unit, simply by controlling the transmitted power according to the automatic gain control (AGC) measurement of the forward link power received by the mobile receiver. Generally, however, this is not effective enough: the forward and reverse link propagation losses are not symmetric, particularly when their center frequencies are widely separated from one another. Thus, even after adjustment using "open loop" power control based on AGC, the reverse link transmitted power may differ by several decibels from one subscriber to the next.

The remedy is "closed loop" power control. This means that when the base station determines that any user's received signal on the reverse link has too high or too low a power level (or more precisely E_s/I_0 level), a one-bit command is sent by the base station to the user over the forward link to command it to lower or raise its relative power by a fixed amount, Δ dB. This creates a so-called "bang-bang" control loop, whose delay is the time required to send the command and execute the change in the user's transmitter. Before considering the control loop performance, we must establish how the decision is made to raise or lower power. As shown in the lower portion of Figure 4.7a, the sum of S successive maximum-of-M correlator outputs is compared with a preset threshold in the same general manner as in the search for new fingers, described in Section 4.6 and shown in the upper portion of Figure 4.7a. It differs, however, in two ways:

(a) The maxima are determined based on the sum of the L fingers being tracked, rather than the single finger whose existence and time delay was searched in the previous case.

(b) The measurement usually contains a signal component (provided the correct correlator maximum is used). The issue therefore is whether its mean level is too high or too low, rather than whether the signal is present at all.

Proceeding then as in Section 4.6, quantizing the maximum sum-of-correlator outputs, we obtain the probabilities P_d and P_u that a down- or up-command is given, as a function of E_s/I_0, respectively, as

$$P_d(E_s/I_0) = Pr\left[\sum_{r=1}^{S} Q(\hat{z}_r) > \phi \mid E_s/I_0\right],$$

(4.70)

$$P_u(E_s/I_0) = 1 - P_d(E_s/I_0).$$

Here, $\hat{z}_r$ is the maximum of the M sums of L correlator outputs for each of the M Hadamard sequences for the rth successive sequence period (which is J symbols or JN chips long) and ϕ is the threshold level. The result is then derived in the same way as the detection probability expressions of (4.66) through (4.69),

$$P_d(E_s/I_0) = \{[\hat{\pi}_{E_s/I_0}(w)]^S\}_{\phi^+},$$

(4.71)

where

$$\hat{\pi}_{E_s/I_0}(w) = \sum_{k=0}^{7} \hat{Q}_k(E_s/I_0)w^k$$

(4.72)

and

$$\hat{Q}_k(E_s/I_0) = \int_{\hat{\theta}_k}^{\hat{\theta}_{k+1}} d\hat{F}_{E_s/I_0}(z)\, dz = \hat{P}_C(z; E_s/I_0)\hat{P}_I^{M-1}(z)\Big|_{\hat{\theta}_k}^{\hat{\theta}_{k+1}},$$

(4.73)

$$k = 0, 1, 2, \ldots.$$

(4.71) through (4.73) differ from their counterparts in the previous section in the following ways:

1. The distribution functions $\hat{P}_C$ and $\hat{P}_I$ pertain to the sum of the L correlator outputs. These are obtained from (4.44) through (4.46) to be

$$\hat{P}_I(z) = \int_0^z \frac{y^{L-1}\exp(-y)}{(L-1)!}\, dy = 1 - e^{-z}\sum_{k=0}^{L-1} z^k/k!$$

(4.74)

$$\hat{P}_C(z; E_s/I_0) = \begin{cases} 1 - \exp[-z/(1+\overline{\mu})] & \text{(Rayleigh} \\ \quad \times \sum_{k=0}^{L-1}[z/(1+\overline{\mu})]^k/k! & \text{fading),} \\ \int_0^z \left(\frac{y}{L\mu}\right)^{(L-1)/2} & \\ \quad \times e^{-(y+L\mu)}\mathcal{I}_{L-1}(2\sqrt{L\mu y})\, dy & \text{(unfaded),} \end{cases}$$

(4.75)

where $\overline{\mu} = J\overline{E_s}/I_0$, $\mu = JE_s/I_0$.

2. The quantization thresholds, $\hat{\theta}_k$, may be chosen to be different from θ_k, the search quantization thresholds.

We now turn our attention to the performance of the control loop.

4.7.1 Power Control Loop Performance

To begin with, although the base station makes the decision to raise or lower the received power, the up or down command must still be transmitted over the forward link[12] to the mobile unit so that it may increase or decrease its transmitted power level by Δ dB. If this command is received in error, the opposite action will be taken. Hence, given that a forward link *command error* occurs with probability π, the actual probability that the power will be reduced for a received energy-to-interference ratio, E_s/I_0, is

$$P'_d(E_s/I_0) = (1 - \pi)P_d(E_s/I_0) + \pi P_u(E_s/I_0)$$
$$= (1 - 2\pi)P_d(E_s/I_0) + \pi$$

and that it will be increased is

$$P'_u(E_s/I_0) = 1 - P'_d(E_s/I_0) = (1 - \pi) - (1 - 2\pi)P_d(E_s/I_0), \quad (4.76)$$

where $P_d(E_s/I_0)$ is given by (4.71) through (4.75).

The purpose of the control loop is to maintain a desired E_s/I_0 level in an environment where the propagation loss varies significantly, but slowly enough that the control mechanism, including inherent delays, can track the changes. Let the total transmitted energy for the jth power control measurement period, of duration $S J N T_c$ chips, be $T(j)$ dB, and let the propagation loss during this period be $L(j)$ dB. Then the received energy over this measurement period is

$$E(j) = T(j) - L(j) \text{ dB.} \quad (4.77)$$

The command loop, assuming a delay of one measurement period (any integer number of measurement periods could equally well be assumed), causes the transmitted power during the $(j + 1)$th measurement interval to increase or decrease by Δ dB. Thus,

$$T(j + 1) = T(j) + \Delta \cdot C[(E(j - 1))], \quad (4.78)$$

---

[12] The method for incorporating up–down commands in the forward link, as well as a more complete description of the process, will be discussed in Chapter 6. Also, the error probability for commands on the forward link, π, will be considerably greater than for ordinary transmission, because it will be sent uncoded to minimize delay.

where

$$C(E) = \begin{cases} -1 \text{ with probability } P'_d(E_s/I_0) \\ +1 \text{ with probability } 1 - P'_d(E_s/I_0). \end{cases} \qquad (4.79)$$

Combining (4.77) and (4.78), we obtain

$$E(j + 1) = E(j) + \Delta \cdot C[E(j - 1)] - [L(j + 1) - L(j)] \text{ dB.} \qquad (4.80)$$

This is the control equation of a time-discrete first-order loop with one-interval delay and the driving function $[L(j + 1) - L(j)]$.

The propagation loss obviously depends strongly on the distance from the subscriber unit to the base station. However, it also depends on shadowing or blockage by buildings and other objects. As already noted, most of the long-term propagation loss due to distance is adjusted by an open loop correction, based on the AGC of the subscriber unit. Very rapid variations, shorter than about one millisecond, are mostly due to Rayleigh fading phenomena that cannot reasonably be mitigated by power control. This leaves shadowing by objects and long-term fading, such as may be experienced by stationary subscribers. These are of long enough duration to be mitigated by the power control loop. Propagation measurements have generally led to modeling these effects by log-normal distributions, which is equivalent to a Gaussian distribution of the decibel values of the loss. However, the loop equation (4.80) requires a model for the first difference of the propagation loss, $L(j + 1) - L(j)$. Clearly, if the loss in decibels has a Gaussian distribution, so will the difference.

It remains only to specify the second-order statistics. We may take the first difference to have zero mean (since an increase in loss is as likely as a decrease) and an arbitrary variance V. For simplicity, we assume that successive first differences are uncorrelated—i.e., that the loss is an *independent increment* process. This assumption appears pessimistic, because such a process requires a wider tracking bandwidth (a larger gain, Δ) than a process with correlated increments. Equation (4.80) is not easily solvable in its present form. We could use simplifying approximations (either by making the increment values discrete and using Markov state methods, or by considering the time-continuous limit and using Fokker–Planck techniques). But it is more useful to simulate (4.80), and from this develop the histograms of E, properly normalized to yield E_s/I_0. These are shown in Figure 4.11 for the power correction step size equal to 0.5 dB, loss increment standard deviation $\sqrt{V} = 0.5$ dB, $\pi = 0.05, J = 6$, and a measurement period encompassing $S = 6$ Hadamard sequence periods. Cases of unfaded transmission were simulated, with $L = 1$ and of equal average com-

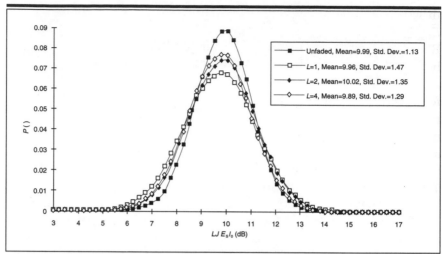

Figure 4.11 Steady-state energy probabilities from simulation. © 1993 IEEE. "Performance of Power-Controlled Wideband Terrestrial Digital Communication" by A. J. Viterbi, A. M. Viterbi, E. Zehavi in *IEEE Transactions on Communications*, Vol. 41, No. 4, pp. 559–569, April 1993.

ponent Rayleigh fading multipath, with $L = 1$, 2, and 4. $C[E(j)]$ was obtained for each value of E_s/N_0, at time j, from (4.79) and (4.71) through (4.76). The threshold ϕ was chosen in each case so that the mean $E/I_0 \approx 10$ dB (where $E = LJE_s$). The histograms appear to be nearly Gaussian with a standard deviation of less than 1.5 dB.

4.7.2 Power Control Implications

The histogram of Figure 4.11 represents the error between the desired $LJ\overline{E}_s/I_0$ and that achieved with power control. But in fact, the required $LJ\overline{E}_s/I_0$ may vary over 2 to 3 dB, since particularly for a mobile user the number of fading paths, and possibly also unfaded paths, may vary. This is clear from Figure 4.8. With a two-way communication system, the base station can monitor the channel status from any subscriber, typically by measuring its error rate. The base station then can vary its closed-loop power control threshold, ϕ for that subscriber. This imposes a higher or lower average $\overline{E}_s/I_0$ on the reverse link of that subscriber, corresponding to its current channel condition.[13] The resulting tighter control can achieve

---

[13] Of course, the absolute $\overline{E}_s/I_0$ required to achieve a given error rate will be strongly influenced by the use of forward error correcting (FEC) coding as treated in the next chapter.

the desired error performance level, but with a larger variation in overall $\overline{E}_s / I_0$. As we shall find in Chapter 6, this variability will affect the overall Erlang capacity of the reverse link, which accounts for the average number of active users supportable on the reverse link to any base station.

We conclude by noting that we have concentrated on the power control for the reverse link, rather than the forward link, for two reasons. First, the power measurement is more complex, since for the forward link, power is measured continuously on the unmodulated pilot signal (see Figure 4.3b and Appendix 4A). More important, power control is critical for achieving maximum reverse link capacity and much less so for the forward link. These topics will be considered in greater detail in Chapter 6.

Chernoff Bound with Imperfect Parameter Estimates

Referring to Equation (4.29) and Figure 4.3, we have, dropping index k,

$$
\begin{aligned}
P_{\mathrm{E}} &< \exp\left\{-\rho N\left[\sqrt{E_{\mathrm{c}}}\sum_{\ell=1}^{L}\alpha_{\ell}\hat{\alpha}_{\ell}\cos(\phi-\hat{\phi}_{\ell})-\rho\sum_{\ell=1}^{L}\hat{\alpha}_{\ell}^{2}I_{0}/4\right]\right\} \\
&= \exp\left\{-\rho N\left[\sqrt{E_{\mathrm{c}}}\sum_{\ell=1}^{L}(\zeta_{\ell}\hat{\zeta}_{\ell}+\eta_{\ell}\hat{\eta}_{\ell})-\rho(I_{0}/4)\sum_{\ell=1}^{L}(\hat{\zeta}_{\ell}^{2}+\hat{\eta}_{\ell}^{2})\right]\right\},
\end{aligned}
$$
$$\rho>0, \quad (4\mathrm{A}.1)$$

where

$$
\begin{array}{ll}
\zeta_{\ell}=\alpha_{\ell}\cos\phi_{\ell}, & \eta_{\ell}=\alpha_{\ell}\sin\phi_{\ell}, \\
\hat{\zeta}_{\ell}=\hat{\alpha}_{\ell}\cos\hat{\phi}_{\ell}\triangleq\zeta_{\ell}+\delta_{\ell}, & \hat{\eta}_{\ell}=\hat{\alpha}_{\ell}\sin\hat{\phi}_{\ell}\triangleq\eta_{\ell}+\varepsilon_{\ell}, \quad (4\mathrm{A}.2) \\
\zeta_{\ell}^{2}+\eta_{\ell}^{2}=\alpha_{\ell}^{2}, & \hat{\zeta}_{\ell}^{2}+\hat{\eta}_{\ell}^{2}=\hat{\alpha}_{\ell}^{2}.
\end{array}
$$

(4A.2) is just a polar-to-rectangular transformation, which is justified by the fact that the rectangular coordinates are actually the parameters being measured. Based on the large number of chips being added to form the estimates, we take $\hat{\zeta}_{\ell}$ and $\hat{\eta}_{\ell}$ to be Gaussian and unbiased with arbitrary variance σ^2. (We shall determine σ^2 for a pilot-aided estimator below.)
Then

$$E(\delta_{\ell}^{2})=E(\varepsilon_{\ell}^{2})=\sigma^{2}, \quad (4\mathrm{A}.3)$$

and thus

$$P_{\mathrm{E}}<\prod_{\ell=1}^{L}\overline{Z}_{\ell}, \quad \text{where}$$

$$
\begin{aligned}
\overline{Z}_{\ell} = \exp\{&-\rho N[\sqrt{E_{\mathrm{c}}}(\alpha_{\ell}^{2}+\zeta_{\ell}\delta_{\ell}+\eta_{\ell}\varepsilon_{\ell})] \\
&+\rho^{2}N[(I_{0}/4)(\alpha_{\ell}^{2}+2\zeta_{\ell}\delta_{\ell}+2\eta_{\ell}\varepsilon_{\ell}+\delta_{\ell}^{2}+\varepsilon_{\ell}^{2})]\}, \quad \rho>0.
\end{aligned}
$$
$$(4\mathrm{A}.4)$$

We choose as before

$$\rho = 2\sqrt{E_c}/I_0. \tag{4A.5}$$

(This is optimum only for the case of perfect estimates, but greatly simplifies the result here without significantly increasing the bound.) Averaging over δ_ℓ and ε_ℓ, we obtain

$$\overline{Z}_\ell = \exp[-N(E_c/I_0)\alpha_\ell^2]$$
$$\times \int_{-\infty}^{\infty} \int_{-\infty}^{\infty} \frac{\exp\{-(\delta_\ell^2 + \varepsilon_\ell^2)[1/(2\sigma^2) - N(E_c/I_0)]\}}{2\pi\,\sigma^2} d\delta_\ell\, d\varepsilon_\ell \tag{4A.6}$$
$$= \frac{\exp[-N(E_c/I_0)\alpha_\ell^2]}{1 - 2\sigma^2 NE_c/I_0}.$$

Now if the estimates of $\hat{\zeta}_\ell = \hat{\alpha}_\ell \cos\hat{\phi}_\ell$ and $\hat{\eta}_\ell = \hat{\alpha}_\ell \sin\hat{\phi}_\ell$ are obtained by averaging N_p chips of the unmodulated pilot, as shown in Figure 4.3,

$$\sigma^2 = \frac{I_0/2}{N_p A_0^2 E_c}. \tag{4A.7}$$

$A_0^2 E_c$ is the energy per chip of the pilot (which is A_0^2 times that of each user). Then from (4A.6) we obtain

$$\overline{Z}_\ell = \frac{\exp[-N(\alpha_\ell^2 E_c/I_0)]}{1 - N/(A_0^2 N_p)}, \qquad N < A_0^2 N_p. \tag{4A.8}$$

Also, since NE_c is the total symbol energy, E_s (assuming $\sum_{\ell=1}^{L} \alpha_\ell^2 = 1$) and $A_0^2 N_p E_c$ is the total pilot energy, we may also express this as

$$\overline{Z}_\ell = \frac{\exp(-\alpha_\ell^2 E_s/I_0)}{1 - (\text{symbol energy/pilot energy})}. \tag{4A.9}$$

Coding and Interleaving

5.1 Purpose

By its very nature, direct sequence spread spectrum CDMA provides a considerably higher dimensionality than needed to transmit information by any single user. This is reflected by the high processing gain, or bandwidth-to-data rate, W/R. In this chapter, we demonstrate how this excess dimensionality or redundancy can be exploited to improve performance, without compromising the other advantages of the high processing gain. Two processing techniques are shown to produce improvements: interleaving and forward error-correcting coding.

5.2 Interleaving to Achieve Diversity

The immediate advantage of excess redundancy is that it provides additional *independent* channel outputs. In Section 4.4.2 we found that, with coherent demodulation, the greater the number of independent path components available in the presence of Rayleigh fading, the better the performance. This is highlighted in Figure 4.4, which shows the total received energy (from all paths combined) required to achieve a given level of performance, as established by the exponent of the Chernoff bound (4.40). The figure shows that with more independent paths, L, less excess total energy is required to achieve a given error exponent for the fading channel. On the other hand, unfaded performance is independent of L.

As noted, besides having L paths, we also have N chips per symbol [see (4.24)]. The problem is that successive chips cannot be regarded as independent; in fact, in Chapter 4 we always assumed the amplitude and phase to be constant over N chip times. It is possible, however, to reorder the chips before transmission so that the N chips belonging to the same symbol are no longer transmitted successively. Rather, they are transmitted at large enough intervals that the time variability of the fading process will lead to independent amplitude and phase for each of the N chips.[1] This reordering procedure to achieve time diversity is called *interleaving* and can be performed in a number of ways. Two of these, called "block" and "convolutional" interleaving, are shown in Figures 5.1 and 5.2, respectively. After reception, the demodulated chip outputs are again reordered to put them back in the original order by the inverse process called *deinterleaving*. The deinterleavers for the block and convolutional interleavers are also shown in Figures 5.1 and 5.2. This process obviously introduces delay between the generation of the digital data and its delivery to the receiving user. As can be seen from the figures, the overall delay produced by the convolutional interleaver–deinterleaver is approximately half of the delay of the block interleaver–deinterleaver.

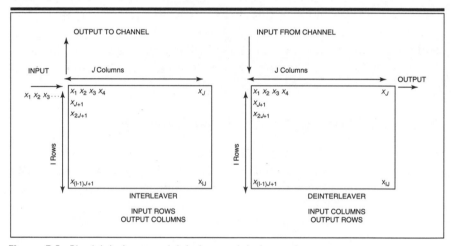

Figure 5.1   Block interleaver–deinterleaver. Interleaved sequence: $x_1, x_{J+1},$ $x_{2J+1}, \ldots , x_{(I-1)J+1}, x_2, x_{J+2}.$ Any two symbols within less than J of each other are at least I apart.

---

[1] Interleaving can be (and usually is) performed not on individual chips, but on subsequences of the N chips, prior to pseudorandom spreading. This reduces memory requirements, since all chips in each sub-sequence, which are kept contiguous after interleaving, have the same sign. Furthermore, with coding, the sub-sequence generally consists of a code symbol. Interleaving coded symbols provides most of the advantage without the excessive memory and processing requirements of chip interleaving. In this section, we carry through the chip interleaving example as a limiting case to demonstrate its power even for uncoded systems, even though it is unrealistic because of the memory requirements and its inapplicability to noncoherent reception.

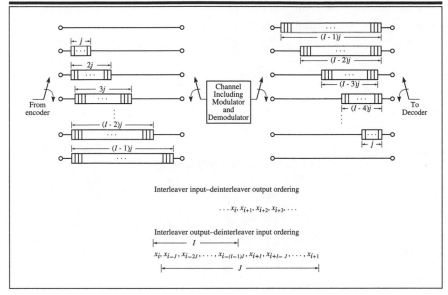

Figure 5.2 Convolutional interleaver–deinterleaver.

Assume that successive chips are interleaved to be far enough apart so that they encounter independent fading effects. Then the impact on performance of a coherent demodulator can be easily determined by modifying Equations (4.27) through (4.30). In particular, (4.28) becomes

$$E[Y_\ell(k) \mid x_n(k) = -1] = -\sum_{n=1}^{N} \sqrt{E_c}\, \alpha_{\ell n}\hat{\alpha}_{\ell n} \cos(\phi_{\ell n} - \hat{\phi}_{\ell n}), \quad (5.1a)$$

$$\mathrm{Var}[Y_\ell(k)] = \sum_{n=1}^{N} \alpha_{\ell n}^2 I_0/4, \quad (5.1b)$$

where $\alpha_{\ell n}$ and $\phi_{\ell n}$ are the amplitude and phase of the nth chip for the ℓth path, and $\hat{\alpha}_{\ell n}$ and $\hat{\phi}_{\ell n}$ are estimates of those values as obtained from the pilot. Proceeding as in (4.29) and (4.30), we obtain the Chernoff bound for perfect amplitude and phase estimates,

$$\begin{aligned}
P_E(k) &< \exp\left[-\sum_{\ell=1}^{L}\sum_{n=1}^{N} \alpha_{\ell n}^2 E_c(k)/I_0\right] \\
&= \prod_{\ell=1}^{L}\prod_{n=1}^{N} \exp[-\alpha_{\ell n}^2 E_c(k)/I_0] \triangleq \prod_{\ell=1}^{L}\prod_{n=1}^{N} Z_{\ell n},
\end{aligned} \quad (5.2)$$

with an expression similar to (4.31) for imperfect estimates.

Interleaving has no impact on unfaded or fixed path amplitudes and phases, for if we normalize the amplitudes such that

$$\sum_{\ell=1}^{L} \sum_{n=1}^{N} \alpha_{\ell n}^{2} = 1,$$

then the first expression in (5.2) is exactly the classical Chernoff bound for error probability in additive Gaussian noise. On the other hand, in *Rayleigh fading*, the independence of the N chips in a given symbol achieved by *interleaving* modifies (4.39) so that it becomes

$$\overline{P}_{\mathrm{E}} < \left[\frac{1}{1 + (\overline{E}_{c}/I_{0})} \right]^{LN} = \exp[-\ln(1/Z)],$$

where

$$\ln(1/Z) = LN \ln[1 + (\overline{E}_{c}/I_{0})]. \tag{5.3}$$

If we plot the excess total energy-to-interference ratio, $LN E_c / I_0 / \ln(1/Z)$ (in all L paths and all N chips) as a function of the exponent $\ln(1/Z)$, we obtain exactly the same curves as in Figure 4.4, except that now the parameter of the curves is LN rather than just L. Since for a given number of independent paths, L, the parameter is now increased by the factor $N > 1$, the curve will lie much closer to the unfaded case. The required energy-to-interference for a given error performance is thus reduced (dramatically for $L = 1$ and significantly even for $L = 3$).

This is just another example of the well-known fact that, in time-varying fading channels, diversity can improve performance markedly. Such diversity can be attained by spatial separation, as with multiple antennas or by multiple paths provided by nature, or by temporal separation through the interleaving process just described. However, the latter comes at the price of delay. The slower the fading process, the larger the necessary interleaving span and hence delay required to achieve independence through time-diversity.

5.3 Forward Error Control Coding—Another Means to Exploit Redundancy

We now turn to a universally effective method of exploiting redundancy, forward error-correcting (or error control) coding (FEC). Unlike interleaving alone, FEC improves performance for fixed amplitude and phase channels as well as for fading channels. We begin by applying FEC to fixed coherently demodulated signals in additive Gaussian noise and interference. We then proceed to combine FEC and interleaving for randomly time-varying fading signals with coherent demodulation. Finally, we consider the more difficult problem of noncoherently demodulated signals for both the fixed and fading cases. The delay introduced by FEC is minimal compared to the interleaving delay described in the last section.

We shall consider only convolutional codes, in keeping with common practice. For a justification based on a performance comparison with block codes, see Viterbi [1971] and Viterbi and Omura [1979], Chapters 4 and 5.

5.3.1 Convolutional Code Structure

An FEC convolutional encoder can be viewed as an additional level of digital linear filtering (over the binary field) which introduces redundancy in the original digital data sequence. Such redundancy is already present in a spread spectrum system and available for exploitation. Figure 5.3 illustrates one of the simplest nontrivial convolutional encoders. It is implemented by a shift register that is remotely related to the shift register generator of pseudorandom sequences described in Section 2.2. However, there are two fundamental differences. First, a new input is provided every shift register clock cycle (as compared to the maximal length sequence generator, in which an initial input vector of finite length produces a periodic sequence of length exponentially related to the generator length). Second, there are *more output symbols than input symbols.* This FEC encoder is a finite-state machine that can be most easily described in terms

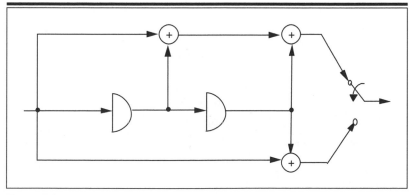

Figure 5.3 Convolutional encoder with $K = 3$, $r = \frac{1}{2}$.

of its state diagram. This is shown in Figure 5.4, where the nodes or states refer to the contents of the register (delay element outputs) just before the next input bit arrives and are so labeled. The input bit "0" or "1" is indicated on the branches, which form the transitions between nodes: Each branch is a *solid* line for an input "0" and a *dotted* line for a "1." The encoder outputs (which consist of two symbols per input bit for the FEC encoder of Figure 5.3) are shown as labels on the transition branches. Thus, for an initial condition (or state) of 00, the input sequence 11010 produces an output sequence 1101010010. This can be seen either from the shift register encoder of Figure 5.3 or its state diagram representation, Figure 5.4.

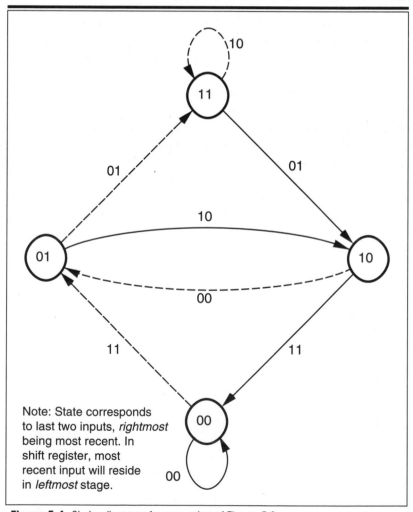

Note: State corresponds to last two inputs, *rightmost* being most recent. In shift register, most recent input will reside in *leftmost* stage.

Figure 5.4 State diagram for encoder of Figure 5.3.

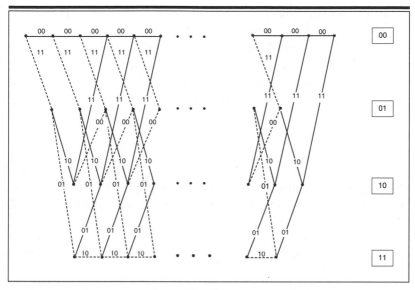

Figure 5.5 Trellis-code representation for encoder of Figure 5.3.

Though it is not really necessary for analyzing either the code characteristics or the performance of the optimal decoder, it is useful in understanding both to exhibit the code on a trellis[2] diagram (Figure 5.5). A trellis diagram is an infinite replication of the state diagram. The nodes (states) at one level in the trellis are reached from the node (states) of the previous level by the transition through one branch, corresponding to one input bit, as determined by the state diagram. Any codeword of a convolutional code corresponds to the symbols along a path (consisting of successive branches) in the trellis diagram. Performance of the coded system depends on the relative *Hamming distance* between codewords: the number of symbols in which they differ. The *"free" distance* is defined as the minimum Hamming distance between any two paths over their unmerged span. We will show that it is the dominant factor in performance evaluation, with error probability decreasing exponentially with increasing free distance.

It is simple, as we shall show, to determine the distance of all paths from the all-zeros (input and output) path over the span over which they are unmerged with the latter. Fortunately, this set of all distances of paths relative to the all-zeros is also the set of all distances from any one path to

---

[2] Trellis is a term, coined by Forney [1970a], which describes a tree in which a branch not only bifurcates into two or more branches but also in which two or more branches can merge into one.

all other paths over their unmerged spans. This symmetry property follows from the fact that a linear (modulo-2) encoder maps the set of all possible binary input sequences onto a closed set of codewords under modulo-2 addition. That is, if the encoder maps input bit sequence u_i onto the sequence x_i and u_j onto x_j, then the sequence $x_i \oplus x_j$, consisting of the pairwise modulo-2 sums of all corresponding components, is also a valid codeword, being generated by the input bit sequence $u_i \oplus u_j$. (Note that sums of any two input sequences are valid input sequences themselves, since any binary sequence can be an input sequence.) It is equally obvious that the distance between any two codeword sequences x_i and x_j is just the *weight* (meaning the number of "1"s) of their modulo-2 sum, $w(x_i \oplus x_j)$.

Now consider the set of all distances from the all-zeros sequence. This is clearly

$$\{w(0 \oplus x_i) \text{ for all } i\} = \{w(x_i) \text{ for all } i\}, \tag{5.4}$$

since any sequence added to the all-zeros remains unchanged. Similarly, the set of all distances from any other specific output sequence x_m is

$$\{w(x_m \oplus x_j) \text{ for all } j\} = \{w(x_i) \text{ for all } i\}. \tag{5.5}$$

This follows from the facts that the modulo-2 sum of any two codewords is another codeword (closed set), and that $x_m \oplus x_j \neq x_m \oplus x_k$ unless $x_j \equiv x_k$, and hence the cardinality of the set is the same as the cardinality of the original codeword set. Thus, we conclude from (5.4) and (5.5) that the set of distances from 0 is the same as the set of distances from any other codeword.

Now from Figure 5.5 it is easy to establish by tracing paths along the trellis that, of the paths which diverge from the all-zeros (top path) at the first (or any) node, exactly one, the one generated by input 100, remerges with the all-zeros after three branches, with an accumulated Hamming distance of 5. Two remerge with an accumulated distance of 6, one after four branches (that generated by input 1100) and one after five branches (that generated by input 10100); and so forth. The same conclusion can be reached by tracing paths along the state diagram of Figure 5.4. In fact, the state diagram is even more convenient for determining the set of all distances, as well as other useful information. Figure 5.6 shows the state diagram labeled with various monomials in the letters δ, β, and λ. In the figure, the 00 node is split in two and shown as both an *initial* and a *final* node. This reflects the fact that we consider all paths diverging from the all-zeros path and remerging to it after some number of transitions. The

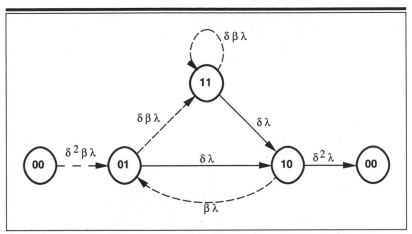

Figure 5.6 State diagram labeled with distance, length, and number of input ones.

monomials labeling the branches indicate the weight of the branch (number of output symbols that are ones) as the exponent of δ, the weight of the input bit (1 for "1" and 0 for "0") as the exponent of β, and the duration in input bits of the branch as the exponent of λ (unity in all cases). Multiplying branch monomials on any traversed path from the initial to the final node yields a monomial with exponents of δ, β, and λ. These are, respectively, the total distance of the codeword sequence from the all-zeros over the unmerged span; the number of inputs in which it differs from the all-zeros input; and the length of the unmerged span. Thus, for example, the monomial products for the three paths traced above are

$$\delta^5\beta\lambda^3, \; \delta^6\beta^2\lambda^4, \text{ and } \delta^6\beta^2\lambda^5.$$

We can obtain a *generating function* for all paths diverging from 0 state at any particular node and remerging at some future time (or by symmetry, remerging with 0 at any particular node and diverging any time previously—allowing the process to start at negatively infinite time). This is the transfer function of the flow graph of Figure 5.6, which is easily computed[3] to be

$$T(\delta, \beta, \lambda) = \frac{\delta^5\beta\lambda^3}{1 - \delta\beta\lambda(1 + \lambda)}. \tag{5.6}$$

---

[3] E.g., either by solving a set of linear equations or by Mason's signal flow graph loop rule [Mason, 1956].

If only the codeword distances and number of input ones are of interest, we may ignore the path length by setting $\lambda = 1$ and obtaining

$$T(\delta, \beta) = T(\delta, \beta, \lambda) \mid_{\lambda=1} = \frac{\delta^5 \beta}{1 - 2\delta\beta}. \qquad (5.7)$$

If only the codeword distances are of interest, we may also set $I = 1$ to obtain

$$T(\delta) = T(\delta, \beta) \mid_{\beta=1} = \frac{\delta^5}{1 - 2\delta}. \qquad (5.8)$$

Expanding this last generating function (by division of polynomials), we obtain

$$T(\delta) = \delta^5 + 2\delta^6 + 4\delta^7 + \cdots + 2^k \delta^{5+k} + \cdots. \qquad (5.9)$$

This means that one path remerges having accumulated distance 5 (differing from the all-zeros in five symbols), two paths with distance 6, and generally 2^k paths with distances $5 + k$. Similarly dividing the polynomials in (5.7), we find

$$T(\delta, \beta) = \delta^5 \beta + 2\delta^6 \beta^2 + 4\delta^7 \beta^3 + \cdots + 2^k \delta^{5+k} \beta^{1+k} + \cdots. \qquad (5.10)$$

This means that generally 2^k paths remerge with accumulated distance $5 + k$, corresponding to an input sequence that contains $1 + k$ ones. This last generating function will prove useful in determining bit error probabilities. It also follows that the exponent of δ in the first term of the expansion is the *free distance* of the code, 5 in this case.

5.3.2 Maximum Likelihood Decoder—Viterbi Algorithm

We have established the codeword structure of convolutional codes by means of the state diagram and trellis diagram. Now we show how easily the same concepts can lead to the optimal decoder for a convolutional code on a memoryless channel—one in which the random impact of the channel on the code symbols is independent from symbol to symbol, possibly achieved through interleaving. Let the input symbol sequence to the channel be denoted $\mathbf{x}$ and the corresponding channel output sequence $\mathbf{y}$. Then as long as the channel is memoryless, the conditional probability, or likelihood function, is given by

$$p(\mathbf{y} \mid \mathbf{x}) = \prod_{\text{all } k} p(y_k \mid x_k). \tag{5.11}$$

The index k ranges over all successive symbols in question in each of the cases to be considered later. However, since time is always bounded, the range of the index can be taken as a finite set of integers. If we seek the most likely path, and hence seek to minimize the sequence error probability over all possible sequences (or paths through the trellis) when all are equiprobable a priori, then we must maximize (5.11) over all input sequences $\mathbf{x}$. The maximum then corresponds to the most likely channel input code sequence, and the corresponding encoder input sequence is the most likely information sequence. Equivalently, we may seek the maximum of the logarithm of (5.11), which is an additive function of the symbol log-likelihoods,

$$\Lambda(\mathbf{x}, \mathbf{y}) = \ln p(\mathbf{y} \mid \mathbf{x}) = \sum_{\text{all } k} \ln p(y_k \mid x_k). \tag{5.12}$$

For reasons that will soon be obvious, it is more convenient to organize (5.12) as the sum of partial sums over branches. We define the *metric* of the jth branch as

$$\mu_j \triangleq \mu(\mathbf{x}_j, \mathbf{y}_j) \triangleq \ln p(\mathbf{y}_j \mid \mathbf{x}_j), \tag{5.13}$$

where the vectors are n-dimensional with n being the number of symbols per branch. Thus,

$$\mu_j = \ln p(\mathbf{y}_j \mid \mathbf{x}_j) = \sum_{k=1}^{n} \ln p(y_{jk} \mid x_{jk}). \tag{5.14}$$

(In the present example, Figures 5.3 to 5.6, $n = 2$, but we shall soon generalize to arbitrary integer values.) Then for the channel input sequence path $\mathbf{x}$ and its output $\mathbf{y}$, the log-likelihood (5.12) can be written

$$\Lambda(\mathbf{x}, \mathbf{y}) = \sum_{j = \text{all branches}} \mu(\mathbf{x}_j, \mathbf{y}_j). \tag{5.15}$$

The maximum likelihood decoder then seeks the path for which the sum of the branch metrics, as defined by (5.13) and (5.14), is maximized. Figures 5.7a and b illustrate two channel examples and the corresponding

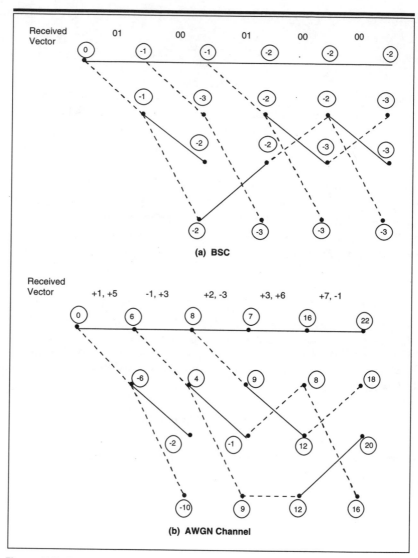

Figure 5.7 Examples of decoding for encoder of Figure 5.3 (decoder state metrics are encirled). (a) BSC. (b) AWGN channel.

decoder for the code of Figures 5.3 under consideration. In each case, the output sequence **y** is shown at the top of the figure. In (a), the channel is a binary symmetric channel (BSC), wherein **y** is a binary sequence that differs from the channel input code sequence **x** in the symbol positions where errors occur. In this case, taking the symbol error probability as

$p < \frac{1}{2}$, the branch metric

$$\mu_j = \mu(\mathbf{x}_j, \mathbf{y}_j)$$
$$= \ln[p^{d_j}(1-p)^{n-d_j}] = d_j \ln\left(\frac{p}{1-p}\right) + n \ln(1-p), \quad (5.16)$$

where $d_j = d(\mathbf{x}_j, \mathbf{y}_j)$ is the Hamming distance between $\mathbf{x}_j$ and $\mathbf{y}_j$ on the jth branch. But the goal is to maximize the sum of branch metrics over the entire path. Thus, we may add an arbitrary constant to each branch and scale the result by an arbitrary positive number, without changing relative rankings. Noting that $\ln[p/(1-p)] < 0$ for $p < \frac{1}{2}$, we may replace the metric by the scaled metric

$$\hat{\mu}_j = -d_j = -d(\mathbf{x}_j, \mathbf{y}_j) \qquad \text{(BSC)} \qquad\qquad (5.17)$$

without changing relative rankings of all paths. In the example of Figure 5.7a, each branch has the metric 0, -1, or -2, depending on whether its code symbols differ from the received code symbols in 0, 1, or 2 positions.

For the example of Figure 5.7b, the output symbol sequence $\mathbf{y}$ is the output of an additive white Gaussian noise (AWGN) channel for which the normalized transmitted sequence $\mathbf{x}$ is $+1$ or -1, according to whether the code symbol was "0" or "1," respectively. Hence, since $x_{jk}^2 = 1$,

$$\mu_j = \ln p(\mathbf{y}_j \mid \mathbf{x}_j) = \ln\left\{ \exp\left[-\sum_{k=1}^n (y_{jk} - \sqrt{E_s}\, x_{jk})^2 / I_0 \right] / (\pi I_0)^{n/2} \right\}$$
$$= \frac{1}{I_0}\left(-\sum_{k=1}^n y_{jk}^2 - nE_s + 2\sqrt{E_s}\sum_{k=1}^n x_{jk}y_{jk} \right) \qquad (5.18)$$
$$- n \ln(\pi I_0)/2,$$

where E_s is the energy per code symbol. The first sum is obviously the same for the jth branch on all paths. Then, again removing common terms and scaling the result by I_0, the branch metric may be replaced by

$$\hat{\mu}_j = \sum_{k=1}^n x_{jk}y_{jk} \qquad \text{(AWGN).} \qquad\qquad (5.19)$$

Each branch for the example of Figure 5.7b is then the inner product of its code symbols (taken as $+1$ or -1) with the received channel outputs (shown at the top of the figure).[4] These examples obviously generalize to any memoryless channel, as defined by (5.11) through (5.15).

Finding the maximum likelihood path thus consists of finding the path for which the sum of its branch metrics (or their modifications) is the greatest over all paths. For L successive input bits, there are L nodes with two branches emanating from each, so by such a "brute force" technique we would need to examine 2^L paths—a hopeless task for even moderate-length input bit sequences. We now show by way of the preceding examples a simple, yet still optimal, method by which the computational effort can grow only linearly with L. This follows from the observation that for any two paths that remerge at a given node, we can exclude the one with the lesser metric sum: It will forever after remain less than that of the other path with which it merges. This leads us to define the state metric at any node i at time k, $M_i(k)$, as the sum of branch metrics up to that state (node) which is the greatest among the branch metric sums of all paths to that node. Thus, we have the recurrence relation

$$M_i(k + 1) = \text{Max}[M_{i'}(k) + m_{i'i}, M_{i''}(k) + m_{i''i}]. \tag{5.20}$$

i' and i'' are the nodes that have permissible transitions to node i in one branch (bit time), and $m_{i'i}$ and $m_{i''i}$ are the branch metrics μ over the transition branches in question.

The operation of (5.20) is illustrated in the examples of Figures 5.7a and 5.7b, with $M_i(k)$ shown as circled values at each node and each level of the trellis. This is the fundamental operation that reduces the search for the maximum likelihood path from 2^L computations for an L-branch path to merely $4L$, in this case. At each state and at each level in the trellis it involves an *add–compare–select* (ACS) operation, where

Add refers to adding each state metric at the preceding level to the two branch metrics of the branches for the allowable transitions.

Compare refers to comparing the pair of such metric sums for paths entering a state (node) at the given level.

---

[4] For compactness of notation, all the branch metrics are taken to be integers, whereas in general they will be real numbers. On the other hand, metrics are integers for the quantized AWGN, which will be treated in Section 5.4.5.

Select refers to selecting the greater of the two and discarding the other. Thus, only the winning branch is preserved at each node, along with the node state metric. If the *two quantities* being compared are *equal, either branch* may be selected, for the probability of erroneous selection will be the same in either case.

Equation (5.20), in all its simplicity, is commonly known as the Viterbi algorithm [1967a]. To implement the decoder corresponding to this algorithm also requires storage of two sets of data. The first is, of course, the path state $M_i(k)$ updated for each successive level k, but requiring only four registers for the running example pursued here. This is called the *metric memory*. Needless to say, if this metric is a real number, as in the AWGN case, it must be quantized to reasonable accuracy. We shall find, however, that in virtually all practical cases the observables are quantized prior to generation of the metric, making both the observables and the metrics discrete variables.[5]

The second set of data to be stored are the selections at each node or state. Called the *path memory*, which again requires four registers of arbitrary length, it stores the selections at each level until a final decision can be made. The way to attain this final decision on all bits is nearly obvious if we force the encoder to return to its initial all-zeros state. This is done merely by inserting two zeros (in our example) at the end of a bit stream of arbitrary duration, B. This procedure, called "tailing off," effectively makes the convolutional code into a block code of length B bits with $B + 2$ branches and $2(B + 2)$ symbols. The final decision is then made by a procedure called "chaining back": Starting with the last node, we trace the decision path backward from the last decision to the first.

However, it is not necessary to tail off the convolutional code into a block code (inserting dummy zeros) in order to make a decision. With very little performance degradation due to suboptimality, after every set of node selections, we may chain back from the state with the highest metric[6] (among the four at that node level) for a sufficient length (say 10

---

[5] To keep the metrics from all drifting upward or downward to values beyond the range of the memory, it is necessary to occasionally normalize by adding or subtracting the same amount from all registers. It is easily shown that the difference between metrics at the same node level is always bounded [Viterbi and Omura, 1979].

[6] We may alternatively choose to chain back from an arbitrary state, with somewhat more degradation. This can be recovered by extending the chain-back to a greater length of branches.

branches). We make a final decision on this bit corresponding to the more likely of the two branches (bits) that emanated from the node to which we have chained back. We then output the bit to which we have chained back and delete it from the path memory. This process, called *memory trunca- tion,* yields almost the same performance as for the tailed-off code, if chain back is long enough. This is because the probability is very low that an incorrect path, unmerged with the correct path over a very long span, maintains a metric higher than that of the correct path. In fact, this proba- bility decreases exponentially with the length of the path, as we shall find in Section 5.4.

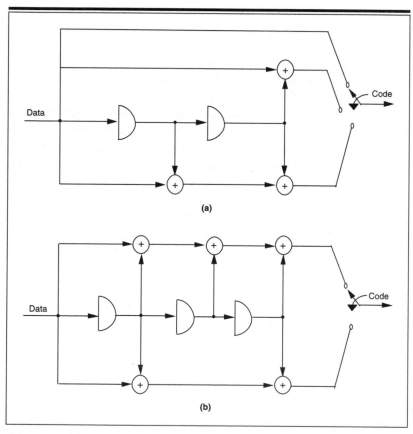

(a)

(b)

Figure 5.8   (a) Convolutional code for $K = 3, r = \frac{1}{2}$. (b) Convolutional code for $K = 4, r = \frac{1}{2}$.

5.3.3 Generalization of the Preceding Example

Two generalizations of the convolutional code example considered up until now should be evident: the length of the encoder shift register, and the number of output taps. The number of stages in the encoder, which is equal to one more than the number of delay elements, is called the *constraint length, K*. Thus, the running example of Figure 5.3 has $K = 3$, while the examples of Figures 5.8a and b have $K = 3$ and $K = 4$, respectively. Also, the running example has two output taps, while that of Figure 5.8a has three. The number of taps determines the convolutional *code rate*, defined as the number of input bits per output symbol. Thus, the running example code rate is $r = \frac{1}{2}$, while that of the example of Figure 5.8a is $r = \frac{1}{3}$. It should then be obvious that all the algorithms and procedures described until now can be applied more generally. The only modifications are that the state diagram and trellis have 2^{K-1} states or nodes (which equals 4 for $K = 3$ and 8 for $K = 4$), and the branches now contain $1/r$ symbols (2 for $r = \frac{1}{2}$; 3 for $r = \frac{1}{3}$).

A slightly less obvious generalization is to a code of rate higher than $\frac{1}{2}$. This can be achieved in two ways. The more direct is by a process called *puncturing*. For example, with a rate $\frac{1}{2}$ code, alternate outputs of the second (lower) tap can be deleted (or punctured)—see Figure 5.9a. Thus, the same number of outputs are taken from the upper tap, but only half as many are taken from the lower tap. The result is a rate $r = \frac{2}{3}$ code, because for each two bits, we generate only three output symbols. This procedure can thus be generalized to delete $n - 1$ out of every $2n$ output symbols from a rate $\frac{1}{2}$ encoder to generate a code of rate $r = n/(n + 1)$, $n \geq 2$. The decoder for a punctured convolutional code is the same as for the original, except that in those positions where the symbol is punctured, and thus is not transmitted, no metric μ is computed. Hence, some branches have a metric based on two symbols, and the others (with punctured symbols) have a metric based on one symbol only.

The most general class of convolutional codes with arbitrary rational rate b/n may be generated by using b parallel registers, each potentially feeding one of n output taps at each stage. An example with $b = 2$ and $n = 3$ is shown in Figure 5.9b. The state diagram for a constraint (shift register) length K will have $2^{b(K-1)}$ states, and each node now has 2^b branches going out and coming into it. The decoder operates as before, except that the path selected is the one with the largest metric among 2^b incoming paths at a node, rather than just the two for rate $1/n$ codes. We shall return to higher-rate codes when we consider performance.

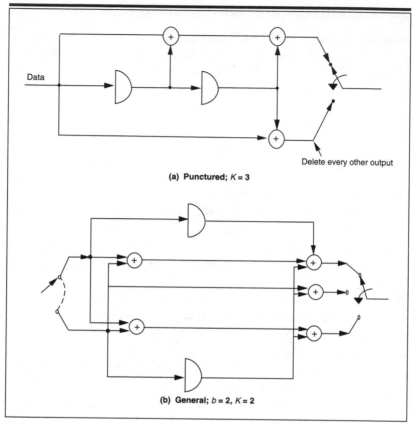

Figure 5.9 Four-state convolutional codes, $r = \frac{2}{3}$. (a) Punctured; $K = 3$. (b) General; $b = 2$, $K = 2$.

5.4 Convolutional Code Performance Evaluation

5.4.1 Error Probability for Tailed-off Block

Evaluation of the error probability of a convolutionally coded system operating over a *memoryless* channel proceeds directly from the generating function derived in Section 5.3 to determine the code distance and other code properties. We continue with the running example of Figure 5.3, noting that generalization to codes with K other than 3 and rates other than $\frac{1}{2}$ follows directly. The generating function of the code's distance properties is given by $T(\delta)$ of Equation (5.8), as described in Section 5.3.1. This enumerates the distance accumulated by all (incorrect) paths unmerged with a given (correct) path over an arbitrary span. By symmetry, this applies both to the set of all possible paths diverging from the correct path at a given node *and* to the set of all possible paths remerging with the correct path at a given node. Taking the latter viewpoint, we can upper-bound the probability of an error in the *selection* step at any given node level by the union bound. The union bound is the sum of the probabilities of error due to the selection of each of the potential incorrect remerging paths. Thus, from the expansion of $T(\delta)$ of (5.9), we find that the probability of a selection error at the jth node is upper-bounded by

$$P_e(j) < P_5 + 2P_6 + 4P_7 + \cdots + 2^k P_{k+5} + \cdots. \qquad (5.21)$$

Here, $P_k = \Pr(\text{pairwise error in favor of an incorrect path that differs in } k$ symbols from the correct path over the unmerged span).

But in Section 4.2 we showed that for a maximum likelihood decoder on a memoryless channel, the Chernoff (or Bhattacharyya) bound on P_1 could be written in the form

$$P_1 < Z, \qquad \text{where } Z = \int_{-\infty}^{\infty} \sqrt{p_0(y)p_1(y)}\, dy. \qquad (5.22)$$

$p_0(y)$ and $p_1(y)$ are the density functions of the channel output conditioned on the input symbol being 0 and 1, respectively, and the integral is replaced by a summation over the values of y if the latter is a discrete rather than a continuous variable. For the AWGN channel, as shown in Section 4.2.1, $Z = e^{-E/I_0}$.

In Chapter 4, we were dealing initially with single symbols, except when the received signal arrived via L multiple paths. In that case [e.g., (4.32)], if the paths were independent, the error probability was the product of terms of the form of (5.22). Here, for a memoryless channel (with uniform statistics), we have d symbols that differ, and hence the Chernoff bound becomes

$$P_d < Z^d. \qquad (5.23)$$

Using (5.22) and (5.23) in (5.21) and comparing with (5.8) and (5.9) yields

$$P_e(j) < Z^5 + 2Z^6 + 4Z^7 + \cdots + 2^k Z^{5+k} + \cdots \qquad (5.24)$$

$$= \frac{Z^5}{1 - 2Z} = T(Z).$$

Suppose now that we are dealing with a "tailed-off" convolutional code constituting a B-bit, $2(B + 2)$-symbol block code, as described in the last section. In this case, since there are B nodes at which a selection is made, the union bound on the error probability of the B-bit block code is

$$P_E < \sum_{j=1}^{B} P_e(j) < BT(Z). \tag{5.25}$$

5.4.2 Bit Error Probability

In contrast with the *block* error probability bound just computed, we may also consider the *bit* error probability of a convolutional code that is not necessarily tailed off into a block. Consider the selection at any given node level (time) j, at the state on the correct path. If the incorrect path is chosen at this time, it will replace the correct path (e.g., all zeros) by an incorrect path for which the input bits differ from it in a certain number of bits. The generating function $T(\delta, \beta)$ accounts for both Hamming distance and the number of input bits that differ between the correct and incorrect paths. It follows from (5.10) that for our running example, the expected number of bit errors caused by a selection error at node level j is

$$E[n_b(j)] \le P_5 + 2(2)P_6 + 4(3)\, P_7 + \cdots + 2^k(k + 1)P_{5+k} + \cdots. \tag{5.26}$$

This is because, of the incorrect paths that potentially remerge with the correct path at any node j, according to (5.10), one remerges at distance 5 from the correct path with one input bit differing, two remerge at distance 6 with two input bits differing, and generally 2^k merge at distance $5 + k$ with $k + 1$ bits differing. The upper bound results because the selection error events are not mutually exclusive. Now using (5.23) in (5.26) and recognizing from (5.10) that the resulting series is merely the derivative[7] of $T(\delta, \beta)$ with respect to β, with β set equal to 1, we find

$$
\begin{aligned}
E[n_b(j)] &< Z^5 + 2(2)Z^6 + \cdots + 2^k(k + 1)Z^{5+K} + \cdots \\
&= \left. \frac{dT(Z, \beta)}{d\beta} \right|_{\beta=1} = \left. \frac{d}{d\beta}\left(\frac{Z^5\beta}{1 - 2Z\beta} \right) \right|_{\beta=1} \\
&= \frac{Z^5}{(1 - 2Z)^2}.
\end{aligned}
\tag{5.27}
$$

---

[7] This is because differentiating the infinite polynomial $T(\delta, \beta)$ with respect to β brings down as a multiplier the β exponent of each term, which is just the number of bit errors for that error event.

To go from this bound on the expected number of bit errors per branch to the bit error probability, it is sufficient to note that the bound on the expected number of bit errors to advance B branches, and hence for B input bits, is bounded by

$$\sum_{j=1}^{B} E[n_b(j)] < B \left. \frac{dT(Z, \beta)}{d\beta} \right|_{\beta=1}. \tag{5.28}$$

Note that some bit errors will eliminate previous bit errors because new (longer) incorrect path selections may supersede incorrect path selections at previous nodes. Nevertheless, the upper bound still holds, because all incorrect path selections at each node are counted whether or not that path ultimately survives.

For arbitrary B, the bit error probability (rate) is just the ratio of the expected number of bit errors to bits decoded (branches advanced). Thus, bit error probability is just (5.28) normalized by B, so that

$$P_b < \left. \frac{dT(Z, \beta)}{d\beta} \right|_{\beta=1}. \tag{5.29}$$

This expression obviously generalizes to any convolutional code of arbitrary constraint length K and rate $1/n$. Note, however, that for rates b/n, where $b > 1$, an advance of B branches means decoding of bB bits. In that case,

$$P_b < \frac{1}{b} \left. \frac{dT(Z, \beta)}{d\beta} \right|_{\beta=1}, \qquad r = b/n. \tag{5.30}$$

Figure 5.10 shows the union–Chernoff bounds on both tailed-off block error probability (with $B = 100$ bits) and bit error probability for the running example of Figure 5.3 as obtained from (5.24) and (5.25) and (5.27) and (5.29), respectively. Error probability is plotted as a function of $\ln(1/Z)/r$ because for the *coherent Gaussian (AWGN) channel,*

$$\ln(1/Z)/r = (E_s/r)/I_0 = E_b/I_0, \tag{5.31}$$

which is the communication system parameter of primary interest.

5.4.3 Generalizations of Error Probability Computation

Generalization of the running example of Figure 5.3 is completely straightforward. Computing the generating function $T(\delta, \beta)$ for state diagrams involving 2^{K-1} states for $K > 5$ is complex, precluding direct alge-

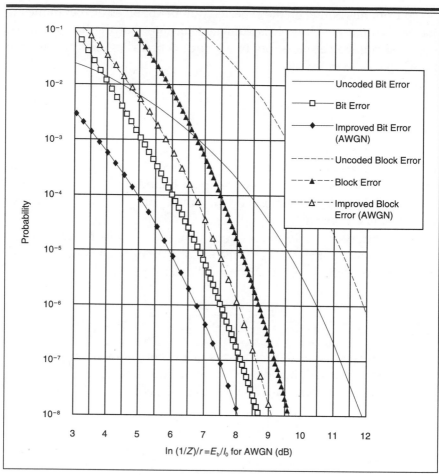

Figure 5.10 Bit error and 100-bit block error probability bounds for $K = 3$, $r = \frac{1}{2}$ convolutional code.

braic manipulation. Several algebraic computation programs exist, but all require too much time at above $K = 7$, at least on moderate-size machines. Fortunately, for a given value of Z, computation of $T(Z)$ can be obtained by numerically solving the state equations:

$$
\begin{bmatrix} x_{00\cdots01} \\ \vdots \\ \vdots \\ x_{11\cdots1} \end{bmatrix} = [\mathbf{A}(Z)] \begin{bmatrix} x_{00\cdots01} \\ \vdots \\ \vdots \\ x_{11\cdots1} \end{bmatrix} + \begin{bmatrix} b_{00\cdots01}(Z) \\ \vdots \\ \vdots \\ b_{11\cdots1}(Z) \end{bmatrix},
$$

$$
T(Z) = [c_{00\cdots01}(Z)\cdots\cdots\cdots c_{11\cdots1}(Z)] \begin{bmatrix} x_{00\cdots01} \\ \vdots \\ \vdots \\ x_{11\cdots1} \end{bmatrix}.
$$

(5.32)

$\mathbf{x}$ is the vector of $2^{K-1} - 1$ (intermediate) state variables, excluding the all-zeros initial and final node. $\mathbf{A}(Z)$ is the matrix of transition branch designators of the state diagram, each term being a monomial in Z. $\mathbf{b}(Z)$ is the vector of designators of those branches connecting the initial node with each intermediate-state node. $\mathbf{c}(Z)$ is the vector of designators of branches connecting each intermediate-state node with the final node. For the example of Figures 5.3 through 5.6, $2^{K-1} - 1 = 3$, and

$$
\mathbf{A}(Z) = \begin{bmatrix} 0 & 1 & 0 \\ Z & 0 & Z \\ Z & 0 & Z \end{bmatrix}, \qquad \mathbf{b}(Z) = \begin{bmatrix} Z^2 \\ 0 \\ 0 \end{bmatrix}, \qquad \mathbf{c}^{\mathrm{T}}(Z) = \begin{bmatrix} 0 \\ Z^2 \\ 0 \end{bmatrix}. \quad (5.33)
$$

Note that for any $r = 1/n$ code, the $\mathbf{A}$ matrix is sparse with only two inputs per row (except the first row, which has a single entry). Also note that the vectors $\mathbf{b}$ and $\mathbf{c}^{\mathrm{T}}$ have only one entry, $\mathbf{b}$ in the first row and $\mathbf{c}^{\mathrm{T}}$ in the row corresponding to $x_{10\ldots0}$. For $r = b/n$ ($b > 1$), the matrix and vectors are less sparse. It also follows from (5.32) that the solution of the state variables can be written

$$
\begin{bmatrix} x_{00\ldots01} \\ \vdots \\ \vdots \\ x_{11\ldots1} \end{bmatrix} = [\mathbf{I} - \mathbf{A}(Z)]^{-1} \begin{bmatrix} b_{00\ldots01}(Z) \\ \vdots \\ \vdots \\ b_{11\ldots1}(Z) \end{bmatrix}, \qquad (5.34)
$$

where $\mathbf{I}$ is the identity matrix. Although the matrix inversion can be performed numerically for any given value of Z, for large K it is easier to perform the expansion

$$
[\mathbf{I} - \mathbf{A}(Z)]^{-1} = \mathbf{I} + \mathbf{A}(Z) + \mathbf{A}^2(Z) + \cdots + \mathbf{A}^k(Z) + \cdots,
$$

truncating after a suitable number of terms. It can be shown that convergence is assured because for any value of the Chernoff bound parameter $Z < 1$, all eigenvalues of $\mathbf{A}(Z)$ are less than unity for noncatastrophic codes, as defined later.

When the bit error probability bound (5.29) or (5.30) is computed in general, the derivative cannot be computed strictly numerically. An arbitrarily close approximation can be obtained, however, through evaluation of $T(Z, 1 + \epsilon)$ and $T(Z, 1 - \epsilon)$ by again solving (5.32). However, in this case matrix $\mathbf{A}$ and vectors $\mathbf{b}$ and $\mathbf{c}$ have components that are the numerical values of the branch monomials in δ and β evaluated at numerical values of Z and $\beta = 1 \pm \epsilon$, respectively. The bit error probability bound of (5.30) is then approximated by

$$P_b \leq \frac{1}{b} \frac{T(Z, 1 + \epsilon) - T(Z, 1 - \epsilon)}{2\epsilon} \tag{5.35}$$

for $r = b/n, b \geq 1$.

It would appear that the smaller ϵ, the better the approximation, although too-small values of ϵ will lead to underflow.

These techniques are applied in Figures 5.11 and 5.12 to computation of the 100-bit block error probability and bit error probability, respectively, for four moderately long codes of constraint length $K = 7$ and

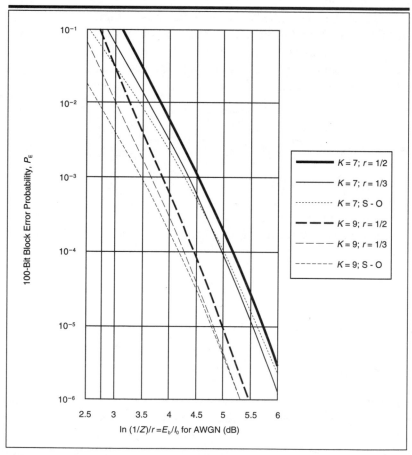

Figure 5.11 100-bit block error probability bounds for various convolutional codes ($K = 7$ and 9).

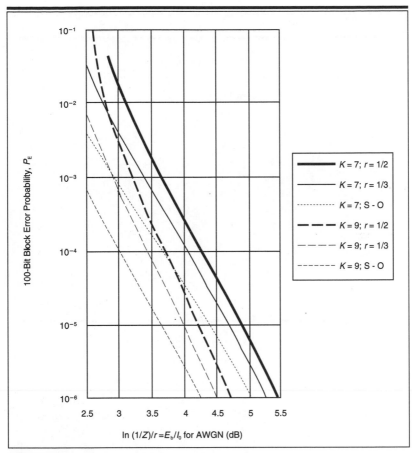

Figure 5.12 Bit error probability bounds for various convolutional codes ($K = 7$ and 9).

$K = 9$, for both $r = \frac{1}{2}$ and $r = \frac{1}{3}$.[8] As for the $K = 3$ code (Figure 5.10), the error probabilities are shown as functions of $\ln(1/Z)/r$ in each case, which for the coherent Gaussian channel equals E_b/I_0 as shown by (5.31). Table 5.1 summarizes the best codes, for $r = \frac{1}{2}$ and $r = \frac{1}{3}$ for $K = 3$ to 10, in the sense of minimum error probability at high E_b/I_0. The tap generator polynomials and free distance are indicated in each case. The $K = 7$ and 9 codes are those whose performance is shown in Figures 5.11 and 5.12.

---

[8] The curves denoted S-O, for superorthogonal codes, will be treated in Section 5.5.

Table 5.1 Minimum Free Distance Convolutional Codes for $K = 3-10$, $r = \frac{1}{2}$ and $\frac{1}{3}$

(A) Rate $\frac{1}{2}$ Codes				
Constraint Length K	Tap Generators in Octal Notation[a]		d_{free}	d_{free} Bound from (5.55)
	G_1	G_2		
3	7	5	5	5
4	17	15	6	6
5	35	23	7	8
6	75	53	8	8
7	171	133	10	10
8	371	247	10	11
9	753	561	12	12
10	1545	1167	12	13

(B) Rate $\frac{1}{3}$ Codes					
Constraint Length K	Tap Generators in Octal Notation			d_{free}	d_{free} Bound from (5.55)
	G_1	G_2	G_3		
3	7	7	5	8	8
4	17	15	13	10	10
5	37	33	25	12	12
6	75	53	47	13	13
7	171	165	133	15	15
8	367	331	225	16	16
9	711	663	557	18	18
10	1633	1365	1117	20	20

[a] E.g., for $K = 3$ and delay operator, D.

$$G_1(D) = 1 \oplus D \oplus D^2,$$
$$G_2(D) = 1 \oplus D^2.$$

5.4.4 Catastrophic Codes

Before leaving the subject of choosing good convolutional codes, we must also recognize and exclude choices that lead to very bad code performance. One such example is the $K = 3$, $r = \frac{1}{2}$ code shown in Figure 5.13. As shown by the encoder in (a), and its state diagram in (b), the input data sequence starting in the 00 state and consisting of all ones of arbitrary length followed by two zeros, 1111 . . . 100, generates the code sequence 11010000 . . . 001101. Since the all-zeros data sequence generates the all-zeros code sequence, the distance between the all-zeros and the tailed-off all-ones (data) path is just 6, no matter how long the data sequence. Thus, over a BSC, a mere four channel errors will cause an arbitrarily long error sequence in the output bits. On an AWGN channel, a perverse noise sequence over just six symbol times will likewise cause an arbitrarily long output error sequence. Such a code is called *catastrophic*.

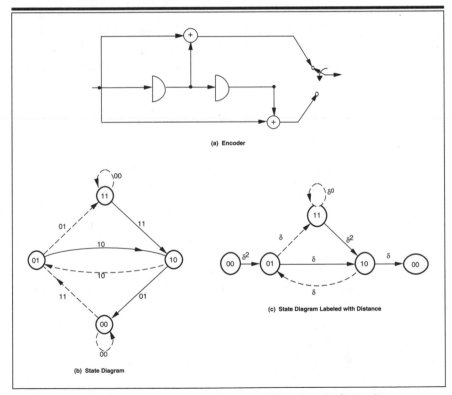

Figure 5.13 Example of a catastrophic code. (a) Encoder. (b) State diagram. (c) State diagram labeled with distance.

A necessary and sufficient condition for a catastrophic code is that a closed loop in the state diagram contains only zeros, or in the labeled state diagram accumulates zero weight. Thus, in the simple example of Figure 5.13, the self loop (state 11 to state 11), has all zeros, in (b), and zero weight, in (c). Massey and Sain [1968] devised a very simple test for whether any code of rate $1/n$ is catastrophic. A necessary and sufficient condition for a code of rate $1/n$ to be catastrophic is that its n generator polynomials (tap sequences written as a polynomial in the delay operator D using modulo-2 addition of terms) have a common factor (modulo-2). Thus, for the present example, the two generator polynomials are $G_1(D) = 1 \oplus D$, $G_2(D) = 1 \oplus D^2$, which with modulo-2 addition have the common factor $1 \oplus D$. To see why this condition is sufficient to make the code catastrophic, consider the infinite sequence obtained from a feedback shift register whose taps were given by the common factor polynomial with an initial condition vector 00 . . . 1 (see Chapter 2). Thus, if the sequence generated by this common factor [in this case, $1/(1 \oplus D) = 1111$ . . . 1 . . . .] is used as an input to the convolutional encoder, it will output two finite length sequences out of the two taps. Hence, all subsequent outputs are zeros. In this case, the output sequences for the two taps have generating functions

$$\frac{G_1(D)}{1 \oplus D} = 1 \qquad (100 \ldots 0 \ldots),$$

$$\frac{G_2(D)}{1 \oplus D} = 1 \oplus D \qquad (1100 \ldots 0 \ldots).$$

Thus, the overall output sequence (alternating between the first and second taps) in response to the all-ones input sequence is 110100 . . . as noted earlier. This differs in only three places from the all-zeros. (The additional three ones mentioned earlier were the result of tailing off with two input zeros.)

Generalization to any number of taps, with any common polynomial factor, is completely straightforward. Finally, Forney [1970b] has shown that for rate $1/n$, of all possible tap sequence selections for a given constraint length, only the fraction $1/(2^n - 1)$ are catastrophic. For example, for $r = \frac{1}{2}$, only one-third of the codes are catastrophic, and these are easily excluded using the Massey–Sain condition.

5.4.5 Generalization to Arbitrary Memoryless Channels—Coherent and Noncoherent

For any convolutional code, we may obtain the bit or block error probability for a memoryless channel as a function of Z, or of the monotonic function thereof, $\ln(1/Z)/r$. This was done in Section 5.4.3 for the four codes with $K = 7$ and 9 and $r = \frac{1}{2}$ and $\frac{1}{3}$, shown in Figures 5.11 and 5.12. For the AWGN channel, the abscissa equals E_b/I_0. For any other memoryless channel (one for which successive symbol metrics are mutually independent), Z must be computed as a function of the channel E_s/I_0. In Chapter 4, this was done for the Rayleigh (and Rician) fading channels with coherent demodulation and for unfaded and faded channels with noncoherent demodulation. In fact, Figure 4.4 (for coherent demodulation) and Figure 4.8 (for noncoherent demodulation) are plots of the total excess required symbol energy over all paths to achieve the same performance bound as for a coherent AWGN channel. This equals $(E_b/I_0)/\ln(1/Z)/r$ in decibels, where E_b is the total bit energy over all paths. Since all parameters are shown in decibels, the ordinates of Figures 4.4 and 4.8 are the *decibel difference* between the required E_b/I_0 (in decibels) and $\ln(1/Z)/r$. Hence, we may obtain this incremental amount from Figures 4.4 and 4.8 for a given code rate and the appropriate value of $\ln(1/Z)$, for any of the channels considered in Chapter 4. This amount must be *added to the abscissa* of Figures 5.10, 5.11, and 5.12 to obtain plots of error probability as a function of E_b/I_0. For example, a $K = 9$, $r = \frac{1}{2}$ convolutional code requires less than $E_b/I_0 = \ln(1/Z)/r = 3.2$ dB or $\ln(1/Z) = 0.2$ dB to achieve a 100-bit block error probability (bound) of 10^{-2} for the coherent AWGN channel. From Figure 4.4 we find that with coherent demodulation of $L = 3$ equal-strength Rayleigh fading paths, the required increment is 0.8 dB. Hence, we require less than $E_b/I_0 = 4.0$ dB for this channel. As another example, consider noncoherent demodulation with $M = 64$ orthogonal signals, $L = 2$ equal-strength Rayleigh paths, and the $K = 9$, $r = \frac{1}{3}$ code. To achieve a 100-bit block error probability (bound) of 10^{-2}, we find from Figure 5.11 that the code requires less than $E_b/I_0 = \ln(1/Z)/r = 3.0$ dB or $\ln(1/Z) = -1.8$ dB in a coherent AWGN channel. Then, from Figure 4.8 it follows that the required increment for the noncoherent Rayleigh channel for $L = 2$ paths is 3.9 dB, which means that overall we require less than $E_b/I_0 = 6.9$ dB.

As noted throughout, all error probabilities are bounds, employing two inequalities, the union bound and the Chernoff (or Bhattacharyya) bound. The latter is the looser of the two in the E_b/I_0 region of interest. Appendix

5A shows that, for the AWGN channel, replacing Chernoff bounds by exact expressions (for pairwise error probabilities) reduces the bounds by about an order of magnitude.[9] This is demonstrated in Figure 5.10 for the $K = 3$, $r = \frac{1}{2}$ code. Figures 5.11 and 5.12 show that an order of magnitude reduction in the ordinate corresponds to an abscissa reduction of about 0.5 dB. Reducing all the preceding E_b/I_0 values by this amount generally gives accurate results for coded systems over memoryless channels. This is because, in a coded system, any decision involves the sum of at least K/r independent symbol metrics, enough to justify a central limit theorem approximation.

5.4.6 Error Bounds for Binary-Input, Output-Symmetric Channels with Integer Metrics

All the results of Chapter 4 that we have used thus far assume symbol and branch metrics that are real numbers. In practice, as noted in Section 5.3.2, these real number metrics must be quantized both for computational processing complexity and for storage memory size. Often the metric variable μ is quantized to Q levels and represented by integer values $\pm a_1$, $\pm a_2, \cdots \pm a_{Q/2}$, when Q is even, and by integer values 0, $\pm a_1$, $\pm a_2, \cdots \pm a_{(Q-1)/2}$, when Q is odd. Under the alternate hypotheses that a "0" or "1" (+ or −) was sent, the probability that the quantized metric takes on integer value q is the probability that the unquantized metric lies in the interval (A_{q-1}, A_q). For Q even, these conditional probabilities are given by

$$P_+(q) = \Pr(A_{q-1} \leq \mu < A_q \mid +),$$
$$P_-(q) = \Pr(A_{q-1} \leq \mu < A_q \mid -),$$

$$(5.36)$$

for $q = \pm a_1, \pm a_2, \cdots \pm a_{Q/2}$ and $A_{-(Q/2+1)} = -\infty$, $A_0 = 0$, $A_{Q/2+1} = +\infty$. For Q odd, the levels are similarly defined, but then the quantization interval for $q = 0$ straddles the origin. For binary inputs, the channel can generally be made *output-symmetric* by choosing appropriate quantization

---

[9] Recall also from Section 4.2.4 that under the weak conditions (4.22), we can (alternatively but more generally) reduce the bound by a factor of 2, which reduces the E_b/I_0 required by about 0.2 dB.

levels.[10] The definition of output symmetry (with integer metrics) is

$$P_+(q) = P_-(-q) \begin{cases} q = \pm a_1, \pm a_2, \cdots, \pm a_{Q/2} & \text{for even } Q, \\ q = 0, \pm a_1, \pm a_2, \cdots, \pm a_{(Q-1)/2} & \text{for odd } Q. \end{cases} \quad (5.37)$$

For binary-input, output-symmetric channels, the Chernoff bound on the pairwise error probability is readily obtained for any two codewords whose signs differ in d symbols. Suppose that there are d symbols in which the channel input (code) symbols differ. In d_1 of the symbols the first codeword has a "+" and the second a "−," while in $d_2 = d - d_1$ of the symbols the first codeword has a "−" while the second has a "+." Then if we assume that the first codeword x_1 was sent and let $\hat{\mu}$ be the quantized integer values, the Chernoff bound on the pairwise error probability is

$$P_d(x_1) = \Pr \left(\sum_{n:x_{1n}=+}^{d_1} \hat{\mu}_n - \sum_{n:x_{1n}=-}^{d_2} \hat{\mu}_n < 0 \mid x_1 \right)$$

$$< \left[\sum_{q \in I_q} P_+(q)e^{-pq} \right]^{d_1} \left[\sum_{q \in I_q} P_-(q)e^{+pq} \right]^{d_2}, \quad p > 0. \quad (5.38)$$

The sum is over the integer set $I_q = \{\pm a_1, \cdots, \pm a_{Q/2}\}$ for even Q and $I_q = \{0, \pm a_1, \cdots, \pm a_{(Q-1)/2}\}$ for odd Q. Now using the channel output symmetry condition (5.37), as well as the symmetry in the set I_q, letting $w = e^{-p}$, and noting that $d_1 + d_2 = d$, we may rewrite (5.38) as

$$P_d(x_1) < \left[\sum_{q \in I_q} P_+(q)w^q \right]^{d_1} \left[\sum_{q \in I_q} P_-(q)w^{-q} \right]^{d_2}$$

$$= \left[\sum_{q \in I_q} P_+(q)w^q \right]^{d_1} \left[\sum_{q \in I_q} P_+(-q)w^{-q} \right]^{d_2} \quad (5.39)$$

$$= \left[\sum_{q \in I_q} P_+(q)w^q \right]^{d}, \quad w < 1.$$

Note that the same expression would result if we had assumed x_2 to be sent. Hence $P_d(x_1) = P_d(x_2) = P_d$. Finally, we must minimize the expression with respect to $w < 1$, yielding

$$P_d < Z^d,$$

---

[10] For example, for the AWGN channel it suffices to make the quantization levels symmetric about the origin, which obviously includes the case of uniform quantization.

where

$$Z = \underset{w<1}{\text{Min}} \left[\sum_{q \in I_q} P_+(q)w^q \right]. \tag{5.40}$$

Thus, (5.40) establishes the parameter that determines the performance for a binary-input, output-symmetric integer metric channel. The degradation due to this quantization can often be made very small by choosing Q only moderately large. The AWGN channel, with an even number of equally spaced quantization levels ($a_1 = \Delta/2$, $a_2 = 3\Delta/2$, $a_3 = 5\Delta/2$, $\cdots$) is an example of this, since

$$P_+(q) = \int_{(q-1)\Delta/2}^{(q+1)\Delta/2} \exp[-(z - \sqrt{2E_s/I_0})^2/2] \, dz/\sqrt{2\pi},$$

$$\tag{5.41}$$

$$P_-(q) = \int_{(q-1)\Delta/2}^{(q+1)\Delta/2} \exp[-(z + \sqrt{2E_s/I_0})^2/2] \, dz/\sqrt{2\pi} = P_+(-q).$$

$q = \pm 1, \pm 3, \ldots, \pm(Q-3)$, with the probabilities for $q = +Q-1$ having instead upper limit $+\infty$ and those for $q = -Q+1$ having instead lower limit $-\infty$. Figure 5.14 shows the ratio $(E_s/I_0)/\ln(1/Z)$ in decibels for this quantized AWGN channel for $Q = 2$ levels (hard quantization), $Q = 4$ levels, and $Q = 8$ levels, as computed from (5.40) and (5.41) optimized relative to Δ. As before, this represents the excess energy in decibels required for this degraded channel to perform as well as the unquantized

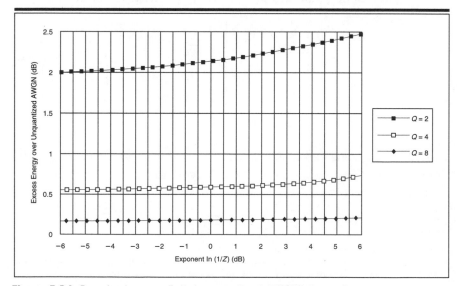

Figure 5.14 Required excess E_b/I_0 for quantized AWGN channels.

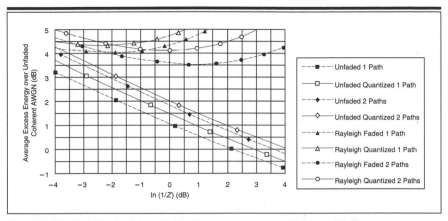

Figure 5.15 Required excess E_b/I_0 (dB) for noncoherent demodulation of orthogonal signals ($M = 64$) with and without quantization.

coherent AWGN channel. Note that the degradation is reduced from 2 dB for $Q = 2$ to 0.2 dB for $Q = 8$. Similar results hold for noncoherent fading channels. As an example, the metric of (4.51), symmetrically quantized to eight levels, requires the excess E_b/I_0 in decibels shown in Figure 5.15. The corresponding unquantized curves of Figure 4.8 are also shown for comparison.

5.5 A Near-Optimal Class of Codes for Coherent Spread Spectrum Multiple Access

5.5.1 Implementation

We now demonstrate that the simple example of Figure 5.3, carried throughout this chapter, can be generalized into a powerful class of codes of arbitrary constraint length K and very low rate, which gives near-optimal performance for coherently demodulated channels with large spreading factors, W/R. Of the three stages of the $K = 3$, $r = \frac{1}{2}$ code of Figure 5.3, the first and last have taps going to both outputs, while the middle stage has only a tap to one output. If we isolate this middle stage, we see that an input "0" to it generates the two output symbols 00, while an input "1" generates 10. These two output symbol pairs form a Hadamard–Walsh block orthogonal set of $M = 2$ codewords.[11]

We now generalize to arbitrary constraint length K, as shown in Figure 5.16a. We again exclude the first and last stages and make the $K - 2$ inner

---

[11] See the first two row and column entries of Figure 4.5 with order or sign inverted, which is immaterial.

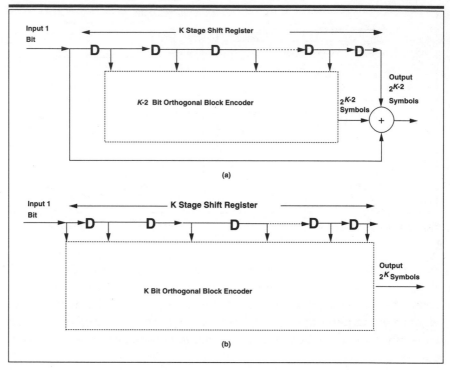

Figure 5.16 (a) Superorthogonal convolutional encoder. (b) Orthogonal convolutional encoder.

stages drive a $K - 2$ bit block orthogonal encoder, whose internal clock runs 2^{K-2} times as fast as the input bit rate (and shifting rate) of the convolutional encoder. Thus, in one input bit time the block orthogonal (Hadamard–Walsh) encoder, as shown in Figure 4.6 with $J = K - 2$, outputs one of 2^{K-2} orthogonal sequences of length 2^{K-2}. For example, for $K = 5$, Figure 4.5 shows the $2^3 = 8$ term sequences prior to the effect of the first and last convolutional encoder stages. Finally, the contents of the first and last stages of the convolutional encoder are both added modulo-2 to every one of the 2^{K-2} symbols of the Hadamard–Walsh sequence. The resulting convolutional code is called a *superorthogonal code* [Zehavi and Viterbi, 1990; Viterbi, 1993a], in contrast with the earlier [Viterbi, 1967b], more conventional orthogonal convolutional code in which the first and last stages also drive the block orthogonal coder (Figure 5.16b). Note that the code rate for the superorthogonal class is $r = 1/2^{K-2}$, while that for the orthogonal class is $r = 1/2^K$. Evidently such codes are usable only with a spread spectrum system, where the code rate r can be as small as the inverse of the spreading factor. For superorthogonal codes, 2^{K-2} can be as large as the spreading factor. It can also be any integer divisor j thereof, in which case each code symbol will consist of j chips of the pseudorandom sequence. For example, if the spreading factor is 128, we may use a $K = 9$ code from this class, assigning one chip per symbol; or a $K = 8$ code, assigning two chips per symbol; or a $K = 7$ code, assigning four chips per symbol; and so forth. For orthogonal convolutional codes, the spreading factor must be at least 2^K. Thus, in this example we can support at most $K = 7$, using one chip per symbol.

5.5.2 Decoder Implementation

The implementation of an efficient maximum likelihood decoder for the encoder of Figure 5.16 is shown in Figure 5.18. It is best understood by examining the metrics to be generated for the case of $K = 5$ ($2^{K-2} = 8$

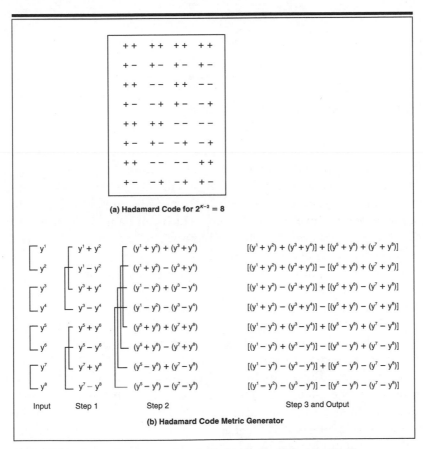

(a) Hadamard Code for $2^{K-2} = 8$

(b) Hadamard Code Metric Generator

Figure 5.17 Example of decoder metric implementation for $K = 5$. (a) Hadamard code for $2^{K-2} = 8$. (b) Hadamard code metric generator.

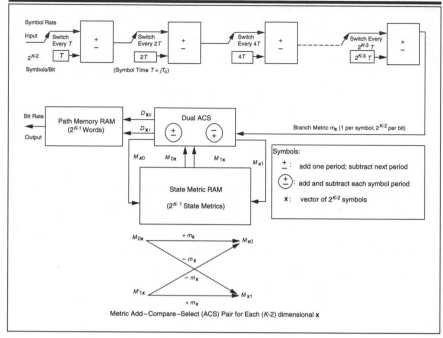

Figure 5.18 Superorthogonal decoder implementation.

orthogonal signals) for the Hadamard–Walsh orthogonal code shown in Figure 5.17a. Given the eight serial BPSK modem outputs $y_1, \ldots, y_8$, the maximum likelihood decoder must generate the eight metrics shown in the third (last) column of Figure 5.17b. As indicated in the figure, these can be generated in $K - 2 = 3$ steps, which can be implemented by the $K - 2$ stage processor at the top of Figure 5.18. Note that this portion of the decoder resembles the central $(K - 2)$ portion of the encoder with mod-2 adders replaced by (real) adders and subtractors, but with the same total number of operations. If this were an orthogonal block code, the 2^{K-2} metrics would then be compared to select the largest. But for the convolutional decoder of Figure 5.18, each metric so generated represents the branch metric to be added to and subtracted from one of the 2^{K-1} state metrics of the convolutional code. Thus, in this case ($K - 2 = 3$), the metric m_x, corresponding to the 3-bit input x, must be added to and subtracted from both the state metrics M_{0x} and M_{1x}. The results must be compared to select the surviving path at each branch (see Section 5.3.2). The add–compare–select (ACS) operations are illustrated by the diagram at the bottom of Figure 5.18, which represents the operation required for each of the $2^{K-2} = 8$ ACS pairs. The remainder of Figure 5.18 shows the implementation of the convolutional (Viterbi) decoder as a serial operation repeated 2^{K-2} times for each ACS pair. At each step, the two updated state metrics are placed back in the appropriate registers of the 2^{K-1} register state metric memory. The decisions are fed to the path memory, which also contains 2^{K-1} registers, one for each of the states, each of length on the order of $5K$. The final decision is arrived at by the standard chain-back algorithm described in Section 5.3.2. Most of the implementation of the ACS and chain-back decisions is standard to all binary convolutional Viterbi decoders, but the nature of the code gives it greater regularity in generating the branch metrics.

It is significant that all operations are performed at the symbol clock speed and, counting the dual ACS as four adders and subtractors, there are precisely $K + 2$ additions or subtractions per symbol time. Thus, the processing complexity (or speed requirement) of the decoder *grows only linearly* with K, although the metric and path memories grow exponentially. Given that the encoder must also perform up to K modulo-2 additions per symbol the decoder's processing complexity is greater than the encoder's only in that real number addition (or subtraction) replaces modulo-2 addition. For spread spectrum (or code division) multiple access, the symbol rate could be no greater than the chip rate, but it could be any integral divisor j thereof. Thus, each symbol contains j chips ($j \geq 1$), and the decoder clock rate for all operations shown is $1/j$th of the pseudorandom sequence shift register clock rate.

5.5.3 Generating Function and Performance

One outstanding feature of this class of codes endows it with superior performance and makes its analysis particularly simple: Every internal branch of the labeled state diagram (generalization of Figure 5.6) has the same weight (distance from the corresponding branch on the all-zeros path or generally between corresponding branches of unmerged paths on the trellis). For $K = 3$ and $r = \frac{1}{2}$, this weight $w = 1$, and in general $w = 2^{K-3} = 1/(2r)$, since two orthogonal sequences differ in exactly half their symbols.

Rather than using the conventional approach of Section 5.3, we obtain the generating function by a simpler method due to McEliece *et al.* [1989]. We consider the set of all paths that diverge from the all-zeros at some point and return to it at any later time, at least K branches later. The distance accumulated is exactly $w = 2^{K-3} = 1/(2r)$ for each branch traversed. Three exceptions are easily noted from Figure 5.16a and shown in Figure 5.19a:

(a) The first branch upon diverging from the all-zeros. In this case there is a single "1" in the first stage, the block code generates all-

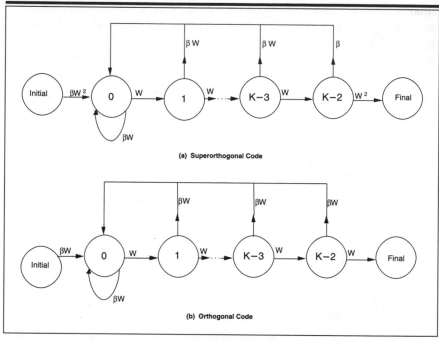

Figure 5.19 Reduced state diagrams for superorthogonal and orthogonal convolutional codes. (a) Superorthogonal code. (b) Orthogonal code.

zeros, and hence the output after mod-2 addition is all ones. Thus, the first branch has weight $2w = 2^{K-2} = 1/r$.

(b) The last branch just before remerging to the all-zeros. The only "1" is in the last stage, and the output of the mod-2 adder is again all ones for a weight $2w = 2^{K-2} = 1/r$.

(c) The branch leading from the state 100 . . . 0 (last before possible remerging) to the state 00 . . . 01 (same as first after diverging). Here, the first and last stages have ones and all the central stages have zeros. In this case the block orthogonal encoder output is all zeros, and it remains that way after each symbol is added mod-2 successively to two ones. *The resulting weight is zero.* (See also Figure 5.4 for $K = 3$.)

Now rather than displaying the preceding information on a conventional 2^{K-1}-state diagram, we may display it on the simple $K + 1$ reduced-state diagram of Figure 5.19. Here we keep track of the number of successive zeros in the input bit stream following the "1" that caused divergence from the all-zeros. Thus, in addition to the initial (diverging) node and the final (remerging) node, we have interior nodes labeled by integers 0 to $K - 2$, corresponding to the number of successive zeros following the "1." It takes $K - 1$ such zeros to remerge by returning to the all-zero state. Thus, the shortest unmerged path is always of length K, corresponding to a "1" followed by $K - 1$ zeros, which is seen to be the direct path from initial to final node. The branches on this path are all of weight w except the first and last, which as noted earlier have weight $2w$. Only the first has an input 1. For notational convenience we denote

$$W \triangleq \delta^w = \delta^{2^{K-3}} = \delta^{1/(2r)}. \tag{5.42}$$

Thus, the first branch on this direct path is labeled βW^2, the last W^2, and all intermediates W. For any interior node, if a "1" is input to the decoder, we must return to the first interior node labeled by 0, indicating no zeros have yet followed a "1." Such a branch in all cases is labeled βW, except the last branch: by exception (c) above, it has zero weight and hence is labeled simply by β. The generating (transfer) function of this graph is easily obtained to be

$$
\begin{aligned}
T_{SO}(W, \beta) &= \frac{\beta W^{K+2}}{1 - \beta(W + W^2 + \cdots + W^{K-3} + 2W^{K-2})} \\
&= \frac{\beta W^{K+2}}{1 - \dfrac{\beta W(1 - W^{K-2})}{1 - W} - \beta W^{K-2}} \\
&= \frac{\beta W^{K+2}(1 - W)}{1 - W[1 + \beta(1 + W^{K-3} - 2W^{K-2})]}.
\end{aligned}
\tag{5.43}
$$

This expression looks unwieldy, but the expressions for error probability are somewhat simplified because the β is set to unity in $T(W, \beta)$ and $dT(W, \beta)/d\beta$. The resulting bounds for B-bit block error probability and for bit error probability are obtained by setting $\beta = 1$ and $\delta = Z$, in the two expressions. This implies by (5.42) that

$$W = Z^{1/2r} = \exp[-\ln(1/Z)/(2r)]. \tag{5.44}$$

Thus, for W given by (5.44),

$$
\begin{aligned}
P_E &< B T_{SO}(W) \\
&= \frac{BW^{K+2}}{1 - 2W}\left(\frac{1 - W}{1 - W^{K-2}}\right)
\end{aligned}
\tag{5.45}
$$

and

$$P_b < \left. \frac{dT_{SO}(W, \beta)}{d\beta} \right|_{\beta=1}$$

$$= \frac{W^{K+2}}{(1 - 2W)^2} \left(\frac{1 - W}{1 - W^{K-2}} \right)^2. \tag{5.46}$$

We can further simplify these bounds by omitting the second factor in each case, noting that since $W < 1$ and $K \geq 3$, $(1 - W)/(1 - W^{K-2}) \leq 1$. [Obviously, for $K = 3$ the ratio is unity and the expressions reduce to (5.24), (5.25), and (5.27)].

Figures 5.20 and 5.21 are plots of the 100-bit tailed-off block error probability and of the bit error probability, obtained respectively from (5.45)

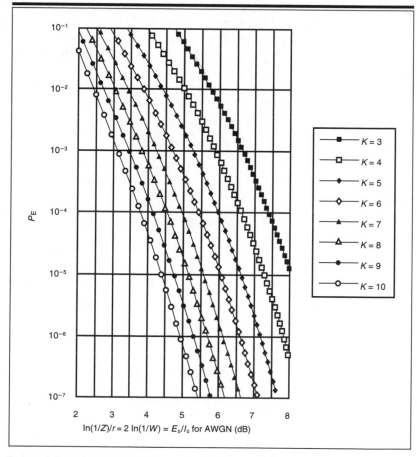

Figure 5.20   100-bit block error probability bounds for superorthogonal codes ($K = 3$ to 10).

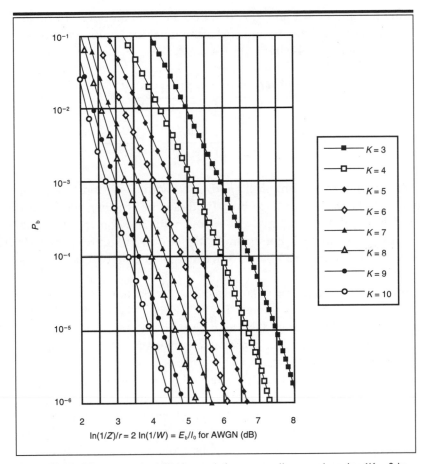

P_b

$\ln(1/Z)/r = 2 \ln(1/W) = E_b/I_0$ for AWGN (dB)

Figure 5.21 Bit error probability bounds for superorthogonal codes ($K = 3$ to 10).

and (5.46), for $K = 3$ through 10, as a function of $\ln(1/Z)/r$, which equals E_b/I_0 for a coherent Gaussian channel.

This entire development applies also to orthogonal convolutional codes (Figure 5.16b). For this lower rate ($r = 2^{-K}$) class, none of the three exceptions apply. Hence, all branches of the reduced-stated diagram (Figure 5.19b) have equal weight $w = 2^{K-1} - 1/(2r)$. Again letting $W = \delta^w$, we obtain from the latter diagram the generating function for an orthogonal convolutional code of constraint length K,

$$T_O(W, \beta) = \frac{\beta W^K(1 - W)}{1 - W[1 + \beta(1 - W^{K-1})]}. \tag{5.47}$$

Letting $\beta = 1$ in both T and $dT/d\beta$, we obtain B-bit block error probability and bit error probability bounds for orthogonal convolutional codes, with W again given by (5.44) but with $r = 1/2^K$:

$$P_E < BT_O(W) = \frac{BW^K(1-W)}{1-2W+W^K} < \frac{BW^K(1-W)}{1-2W}, \qquad (5.48)$$

$$P_b < \left.\frac{dT_O(W,\beta)}{d\beta}\right|_{\beta=1} = \frac{W^K(1-W)^2}{(1-2W+W^K)^2} < \frac{W^K(1-W)^2}{(1-2W)^2}. \qquad (5.49)$$

It is clear from comparing (5.48) and (5.49) with (5.45) and (5.46) that the superorthogonal bounds also hold (with a slight further overbounding) for orthogonal convolutional codes, but with the constraint length reduced from K to $K-2$. For example, the curves in Figures 5.20 and 5.21 for $K = 7$ apply also as bounds for orthogonal convolutional codes for $K = 9$.

For the coherent fading channels studied in Section 4.4, all the bounds apply, but we must increment E_b/I_0 for the AWGN channel by the amount shown in Figure 4.4. Note in particular that for moderately large K, r is very small. Thus, for reasonable values of E_b/I_0, so is $\ln(1/Z)$. In that case the abscissa of Figure 4.4 becomes quite negative in decibels, and the excess E_b/I_0 becomes small.[12] This will not be the case for noncoherent channels (Section 4.5). Making $\ln(1/Z)$ very small, corresponding to very small r, leads to an increase in the excess energy required, primarily because of the noncoherent combining loss. In Section 5.6, we shall consider variations of this approach employing orthogonal convolutional codes that are applicable to noncoherent channels. First, however, we compare the performance of superorthogonal codes with certain limits.

5.5.4 Performance Comparison and Applicability

Figures 5.11 and 5.12 show that superorthogonal codes outperform the best higher-rate ($\frac{1}{2}$ and $\frac{1}{3}$) convolutional codes at low E_b/I_0 values, but by a decreasingly smaller margin for increasingly higher values. At lower values it can be shown that they are asymptotically optimum for the coherent Gaussian channel, as constraint length $K \to \infty$. In fact, as with the less nearly optimal class of orthogonal convolutional codes, their error proba-

---

[12] This assumes, of course, that enough interleaving is provided to make successive code symbols appear independent.

bility approaches zero exponentially with K, as long as $E_b/I_0 > \ln 2$ (-1.6 dB). This is the Shannon limit [Shannon, 1949] for infinite band-width channels.

At high E_b/I_0, however, error probability is dominated by the first term of the union-Chernoff bound expressed in terms of the generating function, $T(W, \beta)$ or its derivative. We now consider how closely optimal performance is approached at high E_b/I_0, or generally, large $\ln(1/Z)/r$. We consider first the additive Gaussian channel without coding, for which

$$\dot{P}_b = Q(\sqrt{2E_b/I_0}) \sim e^{-E_b/I_0}. \tag{5.50}$$

With a superorthogonal code of constraint length K, the leading term of the bit error probability bound is, from (5.46),

$$P_b \sim W^{K+2} = e^{-(K+2)\ln(1/W)}. \tag{5.51}$$

For the Gaussian channel,

$$\begin{aligned} \ln(1/W) &= \ln(1/Z)/(2r) = (E_s/I_0)/2r \\ &= (E_b/I_0)/2. \end{aligned} \tag{5.52}$$

Hence, for large E_b/I_0, where the leading term dominates, we define the ratio of the exponents of coded and uncoded probabilities (5.51) and (5.50) as the *asymptotic coding gain* (ACG). Thus, for superorthogonal convolutional codes,

$$\text{ACG} = \frac{(K+2)\ln(1/W)}{E_b/I_0} = \frac{K+2}{2}, \tag{5.53}$$

while for orthogonal codes it is only $K/2$.

For any convolutional code of rate r and free distance (minimum distance between unmerged codewords) d_f, the leading term of the bit error probability generating function is

$$P_b \sim AZ^{d_f} = Ae^{-d_f \ln(1/Z)}.$$

Here, A is the number of paths remerging with free distance from the correct path at any node, each weighted by the number of bits in which it differs from the correct path. Thus, for the Gaussian channel, ignoring the multiplicative constant A, as K and hence d_f become large,

$$\text{ACG} = \frac{d_f E_s/I_0}{E_b/I_0} = d_f r. \tag{5.54}$$

[More generally, we would have $\ln(1/Z)$ replacing E_s/I_0 and $\ln(1/Z)/r$ replacing E_b/I_0.]

Appendix 5B shows that, for any rate $1/n$ convolutional code, the ACG is upper-bounded by the expression

$$\text{ACG} = d_f r \le \underset{k}{\text{Min}} \frac{(K + k - 1)2^{k-1}}{2^k - 1}. \qquad (5.55)$$

Note, for example, that for $K = 7, r = \frac{1}{2}$, ACG ≤ 5.14. However, for a rate $\frac{1}{2}$ code, ACG can only be an integer or an integer plus one-half, and so in this case ACG ≤ 5 (7 dB). (The upper bound is achieved by the code of Table 5.1 used in Figures 5.11 and 5.12.) On the other hand, the superorthogonal code for $K = 7$ achieves only ACG $= 4.5$ (6.5 dB). Note in Figure 5.11 that for very large values of the abscissa, the superorthogonal code error probability bound crosses over that of the rate $\frac{1}{3}$ code for $K = 7$ and 9. At lower values, however, the small coefficient of the leading and subsequent terms for the superorthogonal code makes its bound considerably lower. It is easily shown by letting $k = 3$ in (5.55), and thus loosening the upper bound, that for any K,

$$\frac{\text{ACG(Superorthogonal)}}{\text{Maximum ACG}} \ge \frac{7}{8}. \qquad (5.56)$$

Thus, the superorthogonal asymptotic coding gain is always within 0.6 dB of the maximum achievable.

Finally, as already noted, superorthogonal codes require very low code rates for even moderate values of K. Thus, for reasonable values of $E_b/I_0 = \ln(1/Z)/r$, $\ln(1/Z)$ is very small or quite negative in decibels. For coherent channels, even with fading (Figure 4.4), the required incremental E_b/I_0 decreases as $\ln(1/Z)$ decreases, as long as symbol interleaving is sufficient. For noncoherent demodulation (Figure 4.8), E_b/I_0 increases as $\ln(1/Z)$ becomes negative in decibels, because of the noncoherent combining loss. Hence, on such channels, performance of this class of codes actually degrades with increasing K. In the next section, we show that with proper modulation, orthogonal convolutional codes will provide improved performance with increasing K even with noncoherent demodulation.

5.6 Orthogonal Convolutional Codes for Noncoherent Demodulation of Rayleigh Fading Signals

5.6.1 Implementation

As just noted, with noncoherent demodulation a very low-rate code is excessively degraded by the loss in noncoherent combining, canceling part or all of its coding gain. To keep this combining loss small to moderate, the energy in each symbol must be on the order of E_b. As in Chapter 4, we may achieve this by selecting one of M orthogonal signals, after aggregating $\log_2 M$ original binary symbols and maintaining coherence over this M-ary symbol time. We now show how this modulation procedure can be coupled with the orthogonal convolutional code of the last section.

The orthogonal convolutional encoder of Figure 5.16b produces for each bit time one of $M = 2^K$ Hadamard–Walsh sequences of length 2^K symbols, $x_1, x_2, \cdots, x_M$. Figure 5.22 shows an example of two paths of the code trellis diagram over their unmerged span, for $K = 6, M = 64$, labeled according to the Hadamard–Walsh sequence generated for each branch. Because the convolutional encoder register contents must be different for the branches of the two paths at corresponding times over the unmerged span, the corresponding output sequences must be different and hence orthogonal. Interleaving is performed on the K-bit blocks that select the

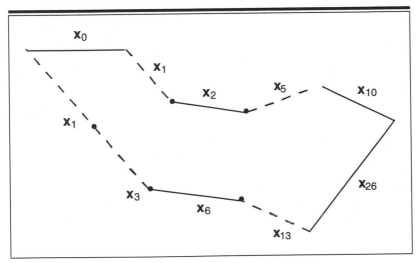

Figure 5.22 Example of two unmerged paths of orthogonal convolutional code.

Hadamard sequences to ensure independence among successive orthogonal signal transmissions.

5.6.2 Performance for L-Path Rayleigh Fading

The noncoherent demodulator correlates over each sequence for each branch for both the I and Q phases. It then forms the sum of the squares. If L equal Rayleigh fading paths are being tracked, the energies are summed to form a branch metric z proportional to the log-likelihood function. This follows from the fact that if the code sequence on the jth branch is $\mathbf{x_n}$, the likelihood of the M orthogonal signal noncoherent demodulator outputs (summed over the L paths) is

$$p_n(z_1, \cdots, z_M) = p(z_n) \prod_{m \neq n}^{M} p(z_m)$$
$$= p_C(z_n) \prod_{m \neq n}^{M} p_1(z_m), \tag{5.57}$$

where $p_C(z)$ and $p_1(z)$ are given by (4.45) and (4.44), respectively. Taking the logarithm of this product yields

$$\ln p_n(z_1, \cdots, z_M) = z_n U/(1 + U)$$
$$+ \sum_{m=1}^{M} [-z_m + (L-1) \ln z_m] + f(U, L). \tag{5.58}$$

Here,[13]

$$U = (\overline{E}_b/I_0)/L \tag{5.59}$$

is the average bit energy-to-interference density per path (noting that the bit energy equals the Hadamard–Walsh symbol energy in this case) and $f(U, L)$ is a function only of U and L. But the second and third terms of (5.58) are independent of n. Hence, to maximize (5.58) we need only preserve the metric z_n for the branch on which the orthogonal code sequence was $\mathbf{x_n}$. Suppose that an incorrect path is unmerged from the correct path for d branches, in which case all corresponding branch pairs are orthogo-

---

[13] Comparing with (4.45), $U = J\overline{E}_s/I_0$. In this case, $J = \log_2 M$ is the number of symbols per bit and $J\overline{E}_s$ is the total received energy per bit in each path. Hence, the total bit energy in all paths, $\overline{E}_b = LJ\overline{E}_s$, and (5.59) follows.

nal. Now let the branch metrics on the correct and incorrect paths be denoted $y_1, y_2, \cdots, y_d$ and $y_1', y_2', \cdots, y_d'$, respectively. Of course, as described earlier, if the transmitted Hadamard–Walsh sequence for the ith branch (of either path) is x_n, then $y_i = z_n$ for that branch. The probability of pairwise error upon remerging is just

$$P_d = \Pr\left(\sum_{i=1}^{d} y_i' > \sum_{i=1}^{d} y_i \right). \qquad (5.60)$$

Independence among successive branch metrics is ensured by interleaving as noted in Section 5.6.1. Independence between metrics of the corresponding branches of the correct and incorrect path is due to orthogonality. We then have the Chernoff bound,

$$P_d < E\left[e^{\rho \sum_{i=1}^{d} (y_i' - y_i)} \right] = \{E[e^{\rho(y' - y)}]\}^d,$$

with

$$
\begin{aligned}
E[e^{\rho(y' - y)}] &= \int_0^\infty e^{\rho y'} p_I(y') \, dy' \int_0^\infty e^{-\rho y} p_C(y) \, dy \\
&= \int_0^\infty \frac{y'^{L-1}}{(L-1)!} e^{(\rho - 1)y'} \, dy' \int_0^\infty \frac{y^{L-1}}{(L-1)!} \frac{e^{-y[\rho + 1/(1+U)]}}{(1+U)^L} \, dy \\
&= \frac{1}{(1-\rho)^L} \frac{1}{[1 + \rho(1+U)]^L}, \qquad \rho > 0. \qquad (5.61)
\end{aligned}
$$

The minimum with respect to ρ occurs at $\rho = (U/2)/(1 + U)$, so that we obtain the Chernoff bound

$$P_d < W_O^d, \qquad (5.62)$$

where

$$W_O = \left[\frac{1 + U}{(1 + U/2)^2} \right]^L, \qquad (5.63)$$

with U defined by (5.59).

The overall B-bit block and bit error probabilities of the constraint length K orthogonal convolutional code are then, according to (5.25) and (5.29),

$$P_E < BT_O(W_O), \tag{5.64}$$

$$P_b < dT_O(W_O, \beta)/d\beta \mid_{\beta=1}. \tag{5.65}$$

$T_O(W, \beta)$ for an orthogonal convolutional code is given by (5.47) and P_E and P_b by (5.48) and (5.49), respectively.

As pointed out in the last section, for a given value of W, the error probability bounds for an orthogonal convolutional code of constraint length K are the same (with a slight overbound) as those of a superorthogonal code of constraint length $K - 2$. Thus, we may use all the curves of Figures 5.20 and 5.21 as performance curves for orthogonal codes for $K = 5$ to 11 with the abscissa taken as $2\ln(1/W_O)$. There remains, however, the excess of E_b/I_0 over $2\ln(1/W_O)$ for this noncoherent L-path Rayleigh channel. Using (5.59) and (5.63), we have the parametric equations[14]

$$\frac{\overline{E}_b/I_0}{2\ln(1/W_O)} = \frac{UL}{2L\ln[(1 + U/2)^2/(1 + U)]} = \frac{U/2}{\ln[(1 + U/2)^2/(1 + U)]},$$

$$\frac{2\ln(1/W_O)}{L} = 2\ln[(1 + U/2)^2/(1 + U)]. \tag{5.66}$$

This excess of $\overline{E}_b/I_0$ over $2\ln(1/W_O)$ in decibels is plotted as a function of $2\ln(1/W_O)/L$ in decibels in Figure 5.23. In conjunction with Figure 5.20 or 5.21, this establishes performance for the noncoherent Rayleigh fading channel. It appears from the latter that $2\ln(1/W_O) = 2$(or 3 dB) is adequate for good performance with an orthogonal code of constraint length 9. But the minimum of Figure 5.23 occurs at approximately $2\ln(1/W_O)/L = 1$ (or 0 dB), and the increment equals 5.3 dB. This requires $L = 2$ paths and achieves $\overline{E}_b/I_0 = 8.3$ dB, which supports using dual (spatial) diversity with Rayleigh fading.[15] Note, however, that even if $L = 1$ so that $2\ln(1/W_O)/L = 2$ (or 3 dB), $\overline{E}_b/I_0$ only increases by 0.6 dB to 8.9 dB. The reason is that with independent successive branches, achieved through interleaving, K-fold (temporal) diversity already exists, and with $L = 2$,

---

[14] These expressions were first obtained by Wozencraft and Jacobs [1965] based on computational cutoff rate considerations.

[15] If dual spatial diversity is not available, the same performance in terms of total $\overline{E}_b/I_0$ can be obtained by repeating each Hadamard–Walsh symbol twice, for half the duration, with interleaving to separate the two repetitions by enough time to provide independence. The foregoing also supports transmitting no more than one bit per Hadamard symbol when interleaving is performed at the Hadamard orthogonal symbol level.

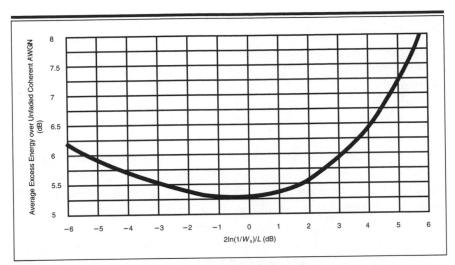

Figure 5.23   Required excess E_b/I_0 for noncoherent Rayleigh fading with orthogonal convolutional codes.

this is merely doubled. Note also that if excessive diversity is available ($L = 4$), the total required bit energy (summed over all paths) only increases marginally by about 0.2 dB. The advantage of diversity is diminished by the loss due to noncoherent combining of lower energy paths. Not surprisingly, too much diversity is less damaging than too little.

Finally, it is worth comparing these results with those of Section 5.4.5. There it was shown that with a $K = 9, r = \frac{1}{3}$ code, which requires the same value of $2\ln(1/W) = 3.0$ dB, the excess was only 3.9 dB for a total $\overline{E}_b/I_0 = 6.9$ dB. The reason for the 1.4 dB difference is the effect of the interleaver in the two cases. This is discussed in the next section.

We note in closing that, for the fixed AWGN channel, diversity does not improve performance, but noncoherent combining still degrades it. Hence, in this ideal case, orthogonal symbol interleaving will be superior to binary symbol interleaving. For obvious reasons, we have concentrated here on the more difficult and usually more practical case of multipath Rayleigh fading, where the foregoing comparisons are more meaningful.

5.6.3 Conclusions and Caveats

We began this chapter by using time interleaving, possibly at the chip level, to provide additional diversity so that fading channels perform almost as well as unfaded channels when coherent demodulation is feasible. We then showed that coding exploits redundancy even for unfaded channels, but the combination of coding and interleaving is most effective for fading channels that are coherently demodulated. We noted, however, that in a coded system, interleaving to the code symbol level is usually sufficient, with code symbols consisting of one to many chips. In the latter case, the interleaving memory is much reduced, but the interleaving delay remains the same.

With noncoherent demodulation, interleaving must be performed at an even coarser level, typically the M-ary orthogonal code symbols, which contain several bits and hence many chips. Otherwise, the additional non-coherent combining loss would counter any coding and diversity gain. In comparing the results of Sections 5.6.2 and 5.4.5 we found that, in pure Rayleigh fading with enough interleaving in both cases to make the successive orthogonal symbols independent, interleaving the binary code symbols before orthogonal symbol selection yielded better performance than interleaving afterwards, using the best convolutional code available for a given K in each case. The reason is that the former provides diversity at the binary code symbol level, which is greater than at the orthogonal symbol level used in the latter. (For example, for $K = 9$ convolutional codes with rate $\frac{1}{3}$ in the former, free distance is 18 binary symbols, while with a $K = 9$ orthogonal convolutional code, free distance is 9 orthogonal symbols.) One must be careful in drawing complexity comparisons, however, because the demodulators are quite different in the two cases and the deinterleaver memory requirements also differ by an order of magnitude, favoring the binary symbol case.

All these comparisons also assume enough interleaving delay to assure the independence of code symbols (binary in one case, M-ary in the other) in close proximity (within a few constraint lengths of one another). Interleaving requires relatively little memory, depending on the rate of fading, but it may cause intolerable delay, particularly when two-way speech is transmitted. If delay constraints limit the interleaver size so that successive or nearby symbols are no longer independent, performance suffers and the relative comparisons may no longer hold. Power control adds to the complexity of the channel model. In such cases, the simple and appealing approach we have demonstrated in this chapter, which allows essentially independent analysis of the code performance and of the channel excess E_b/I_0 requirement over the AWGN channel, no longer applies. While theoretical research on the topic of nonindependent channel coding has continued over several decades [Gallager, 1968; Viterbi and Omura, 1979], no tractable general approach to this difficult problem has emerged. To evaluate and compare modulation, coding, and interleaving techniques for nonmemoryless channels, where successive transmissions are not independent, we must rely primarily on simulations of specific cases.

Improved Bounds for Symmetric Memoryless Channels and the AWGN Channel

In Sections 5.4.1 and 5.4.2, we showed that for generating functions $T(\delta)$ and $T(\delta, \beta)$, the B-bit block error probability and bit error probabilities can be upper-bounded, respectively, by

$$P_E < BT(Z) = B \sum_{k=d_f}^{\infty} a_k Z^k, \qquad (5A.1)$$

$$P_b < \frac{dT(Z, \beta)}{d\beta}\bigg|_{\beta=1} = \sum_{k=d_f}^{\infty} b_k Z^k. \qquad (5A.2)$$

a_k and b_k are the coefficients of the infinite polynomials obtained by dividing the ratio of polynomials that constitute the generating functions. Note, however, that these bounds are the result of applying a union bound on the pairwise error events, followed by a Chernoff bound on the individual event error probabilities. If we can obtain exact expressions for the latter, we can tighten the bounds by applying only the union bound. Thus,

$$P_E < B \sum_{k=d_F}^{\infty} a_k P_k, \qquad (5A.3)$$

$$P_b < \sum_{k=d_F}^{\infty} b_k P_k. \qquad (5A.4)$$

As in (5A.1) and (5A.2), a_k and b_k are the coefficients of the infinite polynomials, derived from the rational functions $T(\delta)$ and $dT(\delta, \beta)/d\beta \,|_{\beta=1}$. P_k is the pairwise error probability between two codewords at Hamming distance k from each other.

For most binary-input, output-symmetric channels, we can derive exact expressions for P_k. Unfortunately we cannot obtain a closed-form expression, as when Chernoff bounds are used in their place. One approach sometimes employed to obtain numerical results is to substitute in (5A.3) and (5A.4) exact expressions for P_k in the first few terms, but then proceed with Chernoff bounds for the remainder (tail) of the sequence of P_k's. We thus obtain a finite sum of terms followed by the tail term in closed form. This bound's value lies between the previous union–Chernoff bound and a union bound of exact pairwise error probabilities.

For the AWGN channel, however, we can obtain a closed-form single-term expression that is nearly as tight as the union bound. In this case, with symbol energy-to-noise density E_s/I_0, the pairwise error for a k-symbol Hamming distance is

$$P_k = Q(\sqrt{2k\,E_s/I_0}) < e^{-kE_s/I_0}, \qquad (5A.5)$$

where

$$Q(z) = \int_z^\infty e^{-\alpha^2/2} d\alpha / \sqrt{2\pi}.$$

Now letting $k = d_f + j, j \geq 0$, we may express (5A.3) and (5A.4) as

$$P_E < B \sum_{j=0}^\infty a_{d_f+j} P_{d_f+j}, \qquad (5A.6)$$

$$P_b < \sum_{j=0}^\infty b_{d_f+j} P_{d_f+j}. \qquad (5A.7)$$

Then from (5A.5) we have

$$P_{d_f+j} = Q(\sqrt{2(E_s/I_0)(d_f + j)}). \qquad (5A.8)$$

We proceed to show that for $x > 0$, $y > 0$,

$$Q(\sqrt{x + y}) < Q(\sqrt{x})e^{-y/2}. \qquad (5A.9)$$

This follows from the following manipulation. Recognizing that for all positive x and y,

$$\sqrt{x+y} - \sqrt{x} \geq 0,$$

$$\sqrt{2\pi}\, Q(\sqrt{x+y}) = \int_{\sqrt{x+y}}^{\infty} e^{-\alpha^2/2} d\alpha = \int_{0}^{\infty} e^{-(\gamma+\sqrt{x+y})^2/2}\, d\gamma$$

$$= \int_{0}^{\infty} e^{-(1/2)(\gamma^2+x+2\gamma\sqrt{x})-(y/2)-\gamma(\sqrt{x+y}-\sqrt{x})} d\gamma$$

$$< \left[\int_{0}^{\infty} e^{-(\gamma+\sqrt{x})^2/2} d\gamma \right] e^{-y/2} = \sqrt{2\pi}\, Q(\sqrt{x}) e^{-y/2},$$

which establishes (5A.9).

Then substituting (5A.9) in (5A.8), (5A.6) becomes

$$P_E < BQ\sqrt{2(E_s/I_0)d_f} \sum_{j=0}^{\infty} a_{d_f+j} e^{-j(E_s/I_0)}$$

$$= BQ\sqrt{2(E_s/I_0)d_f} \sum_{k=d_f}^{\infty} a_k e^{-k(E_s/I_0)} e^{+d_f(E_s/I_0)} \qquad (5A.10)$$

$$= e^{d_f(E_s/I_0)} Q(\sqrt{2(E_s/I_0)d_f})\, BT(Z)\, |_{Z=e^{-E_s/I_0}}.$$

Similarly, (5A.7) becomes

$$P_b = Q(\sqrt{2(E_s/I_0)d_f}) \sum_{k=d_f}^{\infty} b_k e^{-k(E_s/I_0)} e^{+d_f(E_s/I_0)}$$

$$\qquad\qquad\qquad\qquad\qquad\qquad\qquad\qquad (5A.11)$$

$$= e^{d_f(E_s/I_0)} Q(\sqrt{2(E_s/I_0)d_f}) \frac{dT(Z,\beta)}{d\beta} \bigg|_{\beta=1, Z=e^{-E_s/I_0}}.$$

Thus, for the AWGN channel, (5A.10) and (5A.11) require simply a scaling of the union–Chernoff bounds (5.25) and (5.29) by the term $e^{d_f(E_s/I_0)}$ $Q(\sqrt{2(E_s/I_0)d_f}) < 1$. The effect is shown in Figure 5.10 for the $K = 3$, rate $\frac{1}{2}$ code. It applies equally for the AWGN channel for all bounds derived in this chapter.

Upper Bound on Free Distance of Rate 1/*n* Convolutional Codes

We begin by deriving an upper bound due to Plotkin [1960] on the minimum distance between codewords of a linear block code. We then show that it leads easily to the free distance bound for convolutional codes.

The closure property of a linear code ensures that any linear (modulo-2) sum of any two codewords is itself a codeword. Now any nondegenerate binary block code of k bits and $m \geq k$ symbols can be arranged in a matrix of 2^k rows, one for each codeword, and m columns, with no two rows being equal. We show first that any column either consists of all zeros or of 2^{k-1} zeros and 2^{k-1} ones. The proof is by contradiction. Suppose that at least one codeword has a one in the jth column, but that the total number of ones is greater than 2^{k-1} and thus the number of zeros is less than 2^{k-1}. Now if we take one codeword with a one in the jth column and sum it to all other codewords with a one in that column (including itself), we generate more than 2^{k-1} *distinct* codewords with a zero in the jth column, and hence a *contradiction* to the assumption. Similarly, if there were at least one codeword but fewer than 2^{k-1} codewords with a one in the jth column, summing one such codeword with all the codewords having a zero in the jth column would result in more than 2^{k-1} distinct codewords with a one in the jth column—also a *contradiction*. Note, however, that an all-zero column is admissible. In fact, it results from a linear code whose generator (tap sequence) is the all-zeros for the symbols in that position. This proves the proposition.

Suppose then that no column is the all-zeros. Since all columns then must have half zeros and half ones, the average weight of a nonzero codeword is

$$w_{\text{av}} = m2^{k-1}/(2^k - 1).$$

However, if one or more columns consists of all zeros, the average weight will be reduced according to the fraction of all-zero columns so that the average weight is only *upper-bounded* by this expression.

Then the minimum Hamming distance of any linear block code, which equals the minimum nonzero weight, must be less than the average weight of nonzero codewords. Thus,

$$d_{min} \leq w_{av} \leq m2^{k-1}/(2^k - 1), \qquad (5B.1)$$

which is known as the Plotkin bound for linear block codes [Plotkin, 1960].

Now for a binary rate $1/n$, constraint length K, *convolutional code*, consider all codewords that diverge from the all-zeros and remerge within $k + (K - 1)$ branches later, where $k \geq 1$. There are exactly $2^k - 1$ such paths, each containing k bits followed by $K - 1$ zero bits. (If a path remerges earlier, it is kept merged with the correct path through the remainder of the branches.) The set of all such codewords plus the all-zeros constitutes a linear block code of 2^k codewords of length $m = n(k + K - 1)$ symbols. Applying (5B.1), we have that the minimum weight (distance from the all-zeros) of this set is

$$d_{min} \leq n(K - 1 + k)2^{k-1}/(2^k - 1). \qquad (5B.2)$$

Now, considering all possible unmerged lengths, we obtain a bound on the free distance d_f, as the minimum over k of all possible minimum weight unmerged spans. Hence, for any rate $1/n$ code, constraint length K, convolutional code,

$$d_f \leq n \underset{k \geq 1}{\text{Min}} \left[\frac{(K - 1 + k)2^{k-1}}{2^k - 1} \right]. \qquad (5B.3)$$

Here, of course, d_f is at most equal to the largest integer less than or equal to the bound. This bound was first derived by Heller [1968].

Capacity, Coverage, and Control of Spread Spectrum Multiple Access Networks

6.1 General

The treatment of CDMA networks thus far has concentrated almost entirely on the physical layer: the generation, synchronization, modulation, and coding at the transmitter (and the inverse functions at the receiver) of the spread spectrum signals of the multiple access users. Most aspects of these functions are common to all wireless digital communication systems regardless of the multiple access technique. Thus, the fundamentals of Chapters 2 through 5 apply generally, although we have emphasized and used the wideband character of the spread spectrum signals to demonstrate enhanced performance in the presence of interference and multipath propagation. We have relied heavily on these physical layer building blocks. Yet the real advantages of a CDMA network are derived through properly understanding and exploiting the higher-level network concepts and features present in a multicellular multiple access system whose users all fully share common frequency and temporal allocations.

Some of the advantages may appear to be as much qualitative as quantitative, although we shall concentrate on their quantitative evaluation. More importantly, they are heavily intertwined and mutually supportive, as we shall demonstrate throughout this chapter. They can be briefly summarized by the following attributes and techniques, listed roughly in order of importance:

(a) *Universal Frequency Reuse:* This underlies all other attributes. The goal of any wireless communication system is to deliver desired (signal) energy to a designated receiver and to minimize the undesired (interference) energy that it receives. This goal can be achieved by providing disjoint slots (frequency or time) to each user of a cell, or sector of a multi-sectored cell. But once these are fully assigned, the only way to provide more user capacity is by creating more cells. Then users in adjacent cells must also be provided disjoint slots; otherwise, their mutual interference will become intolerable for narrowband (nonspread) transmission. This leads to limited frequency reuse (where typically a slot is reused only once every seventh cell; see Figure 1.1). It also requires frequency plan revision and user channel reallocation every time a new cell is introduced.

With spread spectrum, universal frequency reuse applies not only to all users in the same cell, but also to those in all other cells. Insertion of new cells, as traffic intensity grows, does not require a revision of the frequency plans of existing cells. Equally significant, as more cells are added, the network's ability to insert and extract energy at a given location is enhanced. Hence, the transmitted power levels of both the mobile user and the base station can be reduced significantly by exploiting the power control capabilities of both cells, to be discussed next. In short, since the allocated resource of each user's channel, in both directions, is energy rather than time or frequency, interference control and channel allocations merge into a single approach.

(b) *Power Control:* As just stated, universal frequency reuse demands effective power control of each user, in both directions. Chapter 4 analyzes the physical layer techniques for power control in the reverse (user-to-base station) direction, which is the more critical of the two. In the next section we revisit power control implementation from a more general perspective. We further describe the various algorithms for holding the transmitted power from the mobile users at very near the desired level. In Section 6.6, we evaluate the impact on capacity of power control accuracy. The same is done for the forward (base station-to-user) direction in Section 6.7. Power control ensures that each user receives and transmits just enough energy to properly convey information while interfering with other users no more than necessary. A secondary advantage is that minimizing the transmitted power for a portable user unit maximizes the interval between battery charges.

(c) *Soft Handoff:* Universal frequency reuse makes it possible for a mobile user to receive and send the same call simultaneously from and to two different base stations. In the case of (forward link) reception by the user, the two base station signals can be combined to improve performance, as with multipath combining. In fact, one could regard the second base station signal as a delayed version of the first, generated actively and purposely, rather than as a delayed reflection of the first caused by the environment. For the reverse link, the two different base stations will normally decode the signals independently. Should they decode a given frame or message differently, it will then be up to the switching center (which receives inputs from all base stations and connects the cellular network to the landline switched network) to arbitrate. This issue will be covered in Section 6.3. It is enough to state that the more reliable of the two base stations will generally prevail. Qualitatively, this feature provides more reliable handoff between base stations as a user moves from one cell to the adjacent one. Quantitatively, through proper power control, soft handoff more than doubles capacity of a heavily loaded system. When the network is lightly loaded, it also more than doubles the coverage area of each cell. These measures will be treated quantitatively and in more detail in Sections 6.5 and 6.6.

(d) *Diversity in Space and Time:* These issues were covered in some detail in Chapters 4 and 5, including the constructive combining of multipath components. Section 6.3 shows that soft handoff is another form of spatial diversity, employing the same system implementation. This feature can also be combined with distributed antennas to improve performance, particularly indoors, as discussed in Section 6.8.

(e) *Source Variability:* This includes not only the voice activity first mentioned in Chapter 1, but message or data traffic variability as well. Its effect will be treated in Sections 6.6 and 6.7.

(f) *Antenna Gain:* This applies both to fixed sectorized antennas and to variable pointing by phased arrays. Implementation and limitations will be discussed in Section 6.8.

(g) *FEC Coding Without Overhead Penalty:* As demonstrated in Chapter 5, the inherent excess redundancy due to spectrum spreading can be exploited by low-rate (high redundancy) FEC coding, without the concurrent cost in data rate present in tightly bandlimited channels that allocate frequency or time slots.

In the next two sections, we describe the system implementations of power control and soft handoff, as well as of the pilot signal tracking required for the latter. Quantitative performance measures are addressed in the remainder of the chapter.

6.2 Reverse Link Power Control

We begin by examining the power control of a mobile user's transmitter in a single isolated cell. We assume that the user is already tracking a pilot signal and receiving and transmitting traffic. The means for call initiation by the user are treated in Section 6.6. If the physical channel were completely symmetrical, measuring the power level of the signal received from the base station would determine the transmitter power level required for that user to be received by the base station at the same level as all other users. This ideal situation is almost never the case, since generally, forward and reverse link center frequencies are quite far apart. Nevertheless, suppose we begin by measuring the total power received by the mobile, a function normally performed in receivers by the automatic gain control (AGC) circuit. This signal power coming from the base station is a composite of that destined for all users in the cell or sector. Still it provides a rough measure of the propagation loss to each user, including loss due to range and to shadowing. The higher the received signal power by the user, the lower its transmitted power is set, and vice versa. In fact, the sum of forward and reverse link powers (in decibels) is kept constant by the user terminal's choice of transmitted power. This is called *open-loop power control.*

As already noted, propagation loss is not symmetric. This is primarily because Rayleigh fading caused by near-field component cancellation depends strongly on carrier frequency, which may differ considerably in the two directions. Thus, a *closed-loop power control* mechanism must be included that varies the power transmitted by the mobile, based on measurements made at the base station. In Chapter 4, we described an approach to measuring not power but average E_b/I_0, which is the parameter that most affects error performance. Commands are then transmitted over the forward link to the mobile to raise or lower its transmitted power (by Δ dB) according to whether the E_b/I_0 measured at the base station was lower or higher than desired.[1] The desired E_b/I_0 to be received at the base

---

[1] The power control commands can be multiplexed with the forward link transmission to each user. Because excessive delay in the loop is intolerable, the commands are sent uncoded, thus suffering an error probability that is considerably higher than the forward link coded data. As shown in Section 4.7, however, these errors have a minimal effect on power

station may be estimated a priori to lie between 3 and 7 dB, according to the analyses in Chapters 4 and 5. However, conditions will vary according to the multipath fading environment, particularly if many paths are involved. Thus, as noted in Section 4.7, it is useful to add a power control mechanism called an *outer loop*, which adjusts the desired E_b/I_0 level according to the individual user's error rate measured at the base station. This then guarantees a given error rate per coded voice frame or message packet (typically set at or below 1%). However, the resulting E_b/I_0 parameter, which is already a log-normal random variable— or normal in decibels—because of power control inaccuracy, will have greater variability, manifested as a larger standard deviation of its normally distributed decibel measure. Thus, while Section 4.7 arrives at a typical standard deviation for the closed-loop system between 1.1 and 1.5 dB, the standard deviation caused by the outer loop variations is of the same order of magnitude. Hence, the combination of these two independent components leads to an estimate of total standard deviation on the order of 1.5 to 2.1 dB. It has been measured experimentally [Viterbi and Padovani, 1992; Padovani, 1994] to be between 1.5 and 2.5 dB. We shall assess the effect of variability on capacity in Section 6.6.

6.3 Multiple Cell Pilot Tracking and Soft Handoff

The forward link for each cell or sector generally employs a pilot modulated only by the cell-specific, or sector-specific, pseudorandom sequence, added or multiplexed with the voice or data traffic. This is described in Chapter 4 and shown in Figure 4.2. The pilot provides for time reference and phase and amplitude tracking. It also can be used to identify newly available pilots in adjacent cells or sectors. Specifically, while a user is tracking the pilot of a particular cell, it can be searching for pilots of adjacent cells (using the searching mechanism of its multipath rake receiver). To make this simple and practical, all pilot pseudorandom sequences can use the same maximum length generator sequence, with different initial vectors and hence timing offsets. The relative time-offsets of pilots for neighboring cells and sectors are either known a priori or broadcast to all users of the given cell or sector on a separate

---

control loop performance. One simple method is to puncture (see Section 5.3.3) the forward link code infrequently (e.g., one symbol in 12) and dedicate the punctured symbols to uncoded transmission of commands. The punctured code is then reduced in rate (e.g., by the factor 11/12 if one in 12 symbols is punctured). Its performance is slightly degraded.

CDMA channel, employing its own pseudorandom sequence or time-offset.

Once a new pilot is detected by the searcher and found to have sufficient signal strength (usually relative to the first pilot already being tracked), the mobile will signal this event to its original base station. This in turn will notify the switching center, which enables the second cell's base station to both send and receive the same traffic to and from the given mobile. This process is called *soft handoff*. For forward link transmission to the mobile, the Rake demodulator (Figure 4.3) demodulates both cells' transmission in two fingers of the rake and combines them coherently, with appropriate delay adjustments, just as is done for time-separated multipath components. For the reverse link, normally each base station demodulates and decodes each frame or packet independently. Thus, it is up to the switching center to arbitrate between the two base stations' decoded frames.[2] Soft handoff operation has many advantages. Qualitatively, transition of a mobile between cells is much smoother: The second cell can be brought into use gradually, starting early in the transition of a mobile from one cell to its neighbor cell. Similarly, when the first cell's signal is so weak relative to the second that it cannot be demodulated and decoded correctly, it will be dropped either in response to the mobile's pilot strength measurement or by action of the first cell. Moreover, for any given frame, the better cell's decision will generally be used, with no need to enable a new cell or disable an old one as in classical "hard" handoff. In fact, to avoid frequent handoffs on the boundary between cells (which require excessive control signaling), systems with hard handoff only enable a second cell when its signal strength is considerably above (e.g., 6 dB) that of the first cell. This further degrades performance on the boundary.

Most importantly, however, soft handoff considerably increases both the capacity of a heavily loaded multicellular system and the coverage (area size) of each individual cell in a lightly loaded system. We shall demonstrate this quantitatively, following Viterbi, Viterbi, Gilhousen, and Zehavi [1994]. It is first necessary to determine the mutual interference among cells of a multicellular system.

---

[2] Generally, each frame is provided with an error-detecting code (consisting of a moderate number, c, of check bits at the tail end of the frame) which allows detection of one or more errors with probability on the order of $1-2^{-c}$ [Wolf *et al.*, 1982].

6.4 Other-Cell Interference

6.4.1 Propagation Model

Power control attempts to equalize users' received signal power at a given cell's base station, for all users controlled by that base station. But interference also arrives from users controlled by other cells' base stations. It arrives at the given base station with lower power levels, since soft handoff guarantees that the user is connected at all times through the best base station—the one with the least attenuation due to propagation losses.

The propagation loss is generally modeled as the product of the mth power of distance and a log-normal component representing shadowing losses. This model represents slowly varying losses, even for users in motion, and applies to both reverse and forward links. The more rapidly varying Rayleigh fading losses are not included here.[3] They are already incorporated in the multipath model used in the demodulator analysis (Chapter 4). Thus, for a user at a distance r from a base station, attenuation is proportional to

$$\alpha (r, \zeta) = r^m 10^{\zeta/10}, \tag{6.1}$$

where ζ is the decibel attenuation due to shadowing, with zero mean and standard deviation σ. Alternatively, the losses in decibels are

$$10 \log \alpha (r, \zeta) = 10m \log r + \zeta. \tag{6.2}$$

Experimental data [Jakes, 1974; Lee, 1989] suggest the choices of $m = 4$ for power law and $\sigma = 8$ dB for standard deviation of ζ, the log-normal shadowing.

Any analysis of other-cell interference involves comparison of propagation losses among two or more base stations. Thus, the model must consider the dependence of the propagation losses from a mobile user to two different base stations. Since log-normal shadowing means that the propagation losses in decibels are Gaussian, we assume a joint Gaussian probability density for decibel losses to two or more base stations. Equivalently, we may express the random component of the decibel loss as the

---

[3] However, if the fading is sufficiently slow, as for a mobile traveling at a very slow speed through a region of deep fades, the Rayleigh fading losses may be indistinguishable from shadowing and will be similarly mitigated by power control.

sum of two components: one in the near field of the user that is common to all base stations, and one that pertains solely to the receiving base station and is independent from one base station to another. Thus, we may express the random component of the decibel loss for the ith base station ($i = 0, 1, 2, \ldots$) as

$$\zeta_i = a\xi + b\xi_i, \qquad \text{where } a^2 + b^2 = 1, \qquad a \leq 1, \qquad (6.3)$$

with

$$E(\zeta_i) = E(\xi) = E(\xi_i) = 0,$$

$$\text{Var}(\zeta_i) = \text{Var}(\xi) = \text{Var}(\xi_i) = \sigma^2 \qquad \text{for all } i,$$

$$E(\xi\, \xi_i) = 0 \qquad \text{for all } i,$$

and

$$E(\xi_i\, \xi_j) = 0 \qquad \text{for all } i \neq j.$$

Thus, the normalized covariance (correlation coefficient) of the losses to two base stations, i and j, is

$$E(\zeta_i\, \zeta_j)/\sigma^2 = a^2 = 1 - b^2 \qquad (i \neq j). \qquad (6.4)$$

We may reasonably assume that the near-field and base station–specific propagation uncertainties have equal standard deviations. In that case, $a^2 = b^2 = \frac{1}{2}$ and the normalized covariance is $\frac{1}{2}$ for all pairs of base stations. We shall use these values throughout in all numerical results, although all expressions will be derived for arbitrary covariance value, (6.4).

6.4.2 Single-Cell Reception—Hard Handoff

Suppose first that only a single cell's pilot is being tracked at any one time, and that handoff between cells is performed at the hexagonal cell boundary (Figure 6.1). This is idealized because, as previously noted, even if the boundaries were known, such a process would lead to multiple rapid handoffs for users at or near the boundary. This condition may be alleviated by requiring handoffs to occur only when the second cell's pilot strength is sufficiently above that of the first. Nevertheless, we shall use this idealized hard-handoff model for comparison with the soft handoff results.

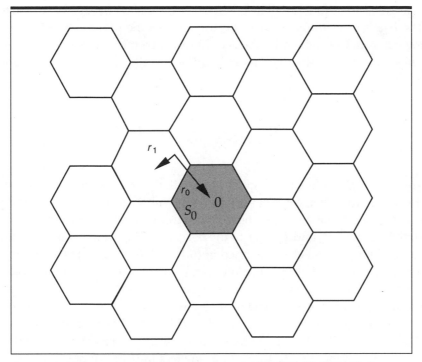

Figure 6.1 S_0 region and relative distances. © 1994 IEEE. ''Soft Handoff Extends CDMA Cell Coverage and Increases Reverse Link Capacity'' by A. J. Viterbi, A. M. Viterbi, K. S. Gilhousen, E. Zehavi, in *IEEE Journal on Selected Areas in Communications,* Vol. 12, No. 8, October, 1994.

We normalize to unity each cell's radius, defined as the maximum distance from any point in the cell to the base station at its center.[4] We also assume a uniform density of users throughout all cells. Letting k_u be the average number of users per cell, then because of the hexagonal shape of the normalized cell, this density is

$$\kappa = \frac{2\,k_u}{3\sqrt{3}}\frac{\text{users}}{\text{unit area}}. \tag{6.5}$$

Then let the distance from the user at coordinates (x, y) to the given cell's base station be $r_0(x, y)$, and that to any other cell's base station be $r_1(x, y)$ (Figure 6.1). Since the user at (x, y) is communicating through the

---

[4] Normalizing cell radius is justified by the fact that all other-cell interference measures are relative and hence normalized by the same-cell interference.

nearest base station, it will also be power-controlled by that base station. The user's transmitter power gain thus equals the propagation loss (6.1) for that cell. Consequently, the relative average interference at the given cell's base station due to *all* users in all other cells, denoted as the region $\overline{S}_0$, is

$$I_{\overline{S}_0} = E \iint_{\overline{S}_0} \left[\frac{r_1^m(x, y)\, 10^{\zeta_1/10}}{r_0^m(x, y)\, 10^{\zeta_0/10}} \right] \kappa\, dA(x, y). \tag{6.6}$$

The subscript 0 refers to the given cell, so that $r_0(x, y)$ is the distance from the user to the given cell's base station. The subscript 1 refers to the cell of occupancy, so that $r_1(x, y)$ is the distance from the user to the *nearest* base station to (x, y). ζ_0 and ζ_1 refer to the corresponding random propagation components in decibels, as defined in (6.1) and (6.2). Thus, the denominator of the bracketed term in the integrand of (6.6) is the propagation loss to the given base station, while the numerator is the gain adjustment through power control by the nearest base station. $\overline{S}_0$ is the entire region outside the given (zeroth) cell, which consists of all other cells. Note also that for each cell in $\overline{S}_0$, $r_1(x, y)$ refers to the distance to a different (nearest) base station. Note finally that all parameters in (6.6) are position-dependent and deterministic except ζ_0 and ζ_1, which are random but do not depend on position. Then, defining

$$R_1(x, y) = r_1(x, y)/r_0(x, y) \qquad \text{and } \beta = \ln (10)/10, \tag{6.7}$$

we may rewrite (6.6) as

$$I_{\overline{S}_0} = E e^{\beta(\zeta_1 - \zeta_0)} \iint_{\overline{S}_0} R_1^m(x, y)\, \kappa\, dA(x, y). \tag{6.8}$$

But from (6.3), we have

$$\zeta_1 - \zeta_0 = b(\xi_1 - \xi_0),$$

which is a Gaussian random variable with zero mean and, since ξ_1 and ξ_0 are independent, $\mathrm{Var}(\xi_1 - \xi_0) = 2\sigma^2$.

Letting $x = \xi_1 - \xi_0$,

$$E e^{\beta b(\xi_1 - \xi_0)} = E e^{\beta b x} = \int_{-\infty}^{\infty} \frac{e^{b\beta x} e^{-x^2/4\sigma^2}}{\sqrt{4\pi}\, \sigma}\, dx = e^{b^2(\beta\sigma)^2}. \tag{6.9}$$

Consequently, from (6.8) and (6.9), using (6.5), we obtain for the mean other-cell interference normalized by the number of users per cell

$$f \triangleq \frac{I_{\bar{S}_0}}{k_u} = e^{b^2(\beta\sigma)^2} \left[\frac{2}{3\sqrt{3}} \iint\limits_{\bar{S}_0} R_1^m(x, y) \, dA(x, y) \right]. \qquad (6.10)$$

The results of numerical integration for $m = 3, 4$, and 5 with $a = b = 1/\sqrt{2}$ are given in Table 6.1.

Table 6.1 Relative Other-Cell Interference Factor, f, for $m = 3, 4$, and 5 and $a^2 = \frac{1}{2}$ with Hard Handoff

Total Standard Deviation, σ	$m = 3$	$m = 4$	$m = 5$
0	0.77	0.44	0.30
2	0.86	0.48	0.33
4	1.18	0.67	0.46
6	2.01	1.13	0.78
8	4.21	2.38	1.64
10	10.9	6.17	4.27
12	35.1	19.8	13.7

6.4.3 Soft Handoff Reception by the Better of the Two Nearest Cells

To approach the performance of a soft handoff system, we next consider other-cell interference when the user is permitted to be in soft handoff to only its two nearest cells. In soft handoff, the user is connected to two or more cell base stations, and instantaneously, on a frame-to-frame basis, the better frame received by either base station is accepted by the network.

Again we consider the zeroth cell. The region for which this cell's base station can be in *soft handoff* with the user, which we again denote S_0, is the six-pointed star that contains the cell (shown as a dotted contour in Figure 6.2). Within S_0, any user that is communicating with one of the six nearest neighbors will introduce interference into the zeroth base station. But this happens only if the propagation loss to that neighbor is less than to the zeroth base station, in which case it is power-controlled by the former. Thus, the mean total interference to the zeroth base station from within

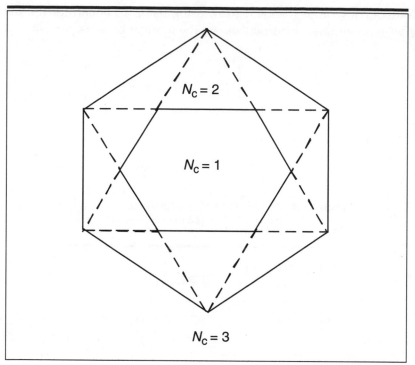

Figure 6.2 S_0 regions for $N_C = 1, 2$, and 3 (within respective boundaries).

the S_0 region is

$$I_{S_0} = \iint\limits_{S_0} R_1^m(x, y) \, E[10^{(\zeta_1 - \zeta_0)/10}; \, r_1^m(x, y) \, 10^{\zeta_1/10} < r_0^m(x, y) \, 10^{\zeta_0/10}] \, \kappa \, dA(x, y),$$

where the expectation is over the sample space for which the inequality is satisfied.

We define

$$M_i(x, y) = 10m \log_{10} r_i(x, y) \tag{6.11}$$

and use the definitions of (6.7) and (6.8). Dropping, for convenience, the notation of dependency on x and y, we obtain

$$I_{S_0} = \iint\limits_{S_0} R_1^m \, E[e^{\beta b(\xi_1 - \xi_0)}; \, \xi_1 - \xi_0 < (M_0 - M_1)/b] \, \kappa \, dA$$

$$= \iint\limits_{S_0} R_1^m \left[\int_{-\infty}^{(M_0 - M_1)/b} e^{\beta bx} \frac{e^{-x^2/4\sigma^2}}{\sqrt{2\pi(2\sigma^2)}} \, dx \right] \kappa \, dA$$

$$= e^{b^2(\beta\sigma)^2} \iint\limits_{S_0} R_1^m \left[\int_{-\infty}^{(M_0 - M_1)/b} \frac{e^{-(x - 2b\beta\sigma^2)^2/4\sigma^2}}{\sqrt{2\pi(2\sigma^2)}} \, dx \right] \kappa \, dA$$

$$= e^{b^2(\beta\sigma)^2} \iint\limits_{S_0} R_1^m \, Q \left[\sqrt{2}b\beta\sigma + \frac{M_1 - M_0}{\sqrt{2}b\sigma} \right] \kappa \, dA,$$

(6.12)

Here,

$$Q(y) = \int_y^\infty e^{-x^2/2} \, / \, \sqrt{2\pi} = \int_{-\infty}^{-y} e^{-x^2/2} \, / \, \sqrt{2\pi},$$

and $R_1(x, y)$ and $M_1(x, y)$, as defined in (6.7) and (6.11), refer to the nearest base station other than the zeroth.

Now, for the complementary region $\overline{S}_0$ (outside the six-pointed star), the two nearest base stations involved in a potential soft handoff do *not* include the zeroth. We let the subscripts 1 and 2 denote the two nearest base stations, which include all contiguous pairs of cells over the entire plane.[5] Then the total mean interference to the zeroth base station from the $\overline{S}_0$ region is

$$I_{\overline{S}_0} = \iint\limits_{\overline{S}_0} R_1^m(x, y) \, E[10^{(\xi_1 - \xi_0)/10};$$

$$r_1^m(x, y) \, 10^{\xi_1/10} < r_2^m(x, y) 10^{\xi_2/10}] \kappa \, dA(x, y)$$

$$+ \iint\limits_{\overline{S}_0} R_2^m(x, y) \, E[10^{(\xi_2 - \xi_0)/10};$$

$$r_2^m(x, y) 10^{\xi_2/10} < r_1^m(x, y) 10^{\xi_1/10}] \, \kappa \, dA(x, y)$$

$$\triangleq I_1 + I_2.$$

(6.13)

---

[5] In numerical computations, only the first three rings of cells are included. The *m*th-power propagation loss renders the contributions from more distant cells negligible.

Again dropping the spatial notation, we evaluate the first integral, using (6.3) and the independence of the ξ_i variables, as follows:

$$
\begin{aligned}
I_1 &= \iint\limits_{S_0} R_1^m E[e^{\beta b(\xi_1 - \xi_0)};\, b\xi_1 + M_1 < b\xi_2 + M_2]\, \kappa\, dA \\[2mm]
&= \iint\limits_{S_0} R_1^m\, E(e^{-\beta b \xi_0}) \left\{ \int_{-\infty}^{\infty} [e^{\beta b \xi_1}\, e^{-\xi_1^2/2\sigma^2}/(\sqrt{2\pi}\,\sigma)] \right. \\[2mm]
&\qquad \left. \times \Pr(b\xi_2 > b\xi_1 + M_1 - M_2)\, d\xi_1 \right\} \kappa\, dA \\[2mm]
&= e^{2b^2(\beta\sigma)^2/2} \iint\limits_{S_0} R_1^m \left[\int_{-\infty}^{\infty} e^{-(\xi_1 - b\beta\sigma^2)^2/2\sigma^2}\, d\xi_1 \right. \\[2mm]
&\qquad \left. \times \int_{\xi_1 + (M_1 - M_2)/b}^{\infty} e^{-\xi_2^2/2\sigma^2}\, d\xi_2/(2\pi\,\sigma^2) \right] \kappa\, dA \\[2mm]
&= e^{b^2(\beta\sigma)^2} \iint\limits_{S_0} R_1^m \left[\int_{-\infty}^{\infty} e^{-x_1^2/2\sigma^2}\, dx_1 \right. \\[2mm]
&\qquad \left. \times \int_{x_1 + b\beta\sigma^2 + (M_1 - M_2)/b}^{\infty} e^{-x_2^2/2\sigma^2}\, dx_2/(2\pi\,\sigma^2) \right] \kappa\, dA \\[2mm]
&= e^{b^2(\beta\sigma)^2} \iint\limits_{S_0} R_1^m\, Q\!\left(\frac{b\beta\sigma}{\sqrt{2}} + \frac{M_1 - M_2}{\sqrt{2}b\sigma} \right) \kappa\, dA.
\end{aligned}
$$

$$(6.14)$$

Now the second integral I_2 is the same as I_1, with M_1 and M_2 interchanged. But since subscripts 1 and 2 represent the two nearest cells, they are interchangeable and hence $I_1 = I_2$. Thus, combining (6.12), (6.13), and (6.14) and using (6.5) and $b = 1/\sqrt{2}$, we obtain the relative interference at the zeroth cell base station from all users not controlled by its base station:

$$
\begin{aligned}
f = \frac{I_{S_0} + I_{\bar{S}_0}}{k_u} = \frac{2e^{(\beta\sigma)^2/2}}{3\sqrt{3}} & \left[\iint\limits_{S_0} R_1^m\, Q\!\left(\beta\sigma + \frac{M_1 - M_0}{\sigma} \right) dA \right. \\[2mm]
& \left. + 2 \iint\limits_{\bar{S}_0} R_1^m\, Q\!\left(\frac{\beta\sigma}{2} + \frac{M_1 - M_2}{\sigma} \right) dA \right].
\end{aligned}
$$

$$(6.15)$$

Again, in the spatial integrals over S_0 and $\bar{S}_0$, R_1, R_2 and M_1, M_2 refer to the two base stations nearest to the user at (x, y).

The value of relative interference, f, is evaluated numerically and shown in Table 6.2. For all m and σ, the value is greatly reduced from the single-cell (hard handoff) case.

Table 6.2 Relative Other-Cell Interference Factor, f, for $m = 3$, 4, and 5 and $\sigma^2 = \frac{1}{2}$ with Two-Cell Soft Handoff

Total Standard Deviation	$m = 3$	$m = 4$	$m = 5$
0	0.77	0.44	0.30
2	0.78	0.43	0.30
4	0.87	0.47	0.31
6	1.09	0.56	0.36
8	1.60	0.77	0.47
10	2.80	1.28	0.73
12	5.93	2.62	1.42

6.4.4 Soft Handoff Reception by the Best of Multiple Cells

Though soft handoff is generally between two cells, it may be among three or more, and in any case, it should always be possible to change the set. We may generalize the results of the last section by increasing the candidate set for handoff to the base stations in the $N_c - 1$ cells nearest to the zeroth cell. Thus, the S_0 region expands. For example, Figure 6.2 shows that S_0 for $N_c = 3$ is a large hexagon that circumscribes the six-pointed star that is S_0 for $N_c = 2$.

Generalizing the analysis of the last section, we find that the mean total interference to the zeroth base station from within the S_0 region is

$$I_{S_0} = \iint_{S_0} \sum_{j=1}^{N_c-1} R_j^m E\left[10^{(\zeta_j - \zeta_0)/10}; \right.$$

$$\left. \operatorname*{Min}_{i=1}^{N_c-1} r_i^m 10^{\zeta_i/10} = r_j^m 10^{\zeta_j/10} < r_0^m 10^{\zeta_0/10} \right] \kappa \, dA$$

$$= \iint_{S_0} \sum_{j=1}^{N_c-1} R_j^m E\left[e^{\beta(\zeta_j - \zeta_0)}; \zeta_0 + M_0 > \zeta_j + M_j = \operatorname*{Min}_{i=1}^{N_c-1} (\zeta_i + M_i) \right] \kappa \, dA$$

$$= \iint_{S_0} \sum_{j=1}^{N_c-1} R_j^m \, E[e^{\beta b(\xi_j - \xi_0)}; \, \xi_i > \xi_j + (M_j - M_i)/b$$

$$\text{for all } i \neq j; \, j \neq 0] \, \kappa \, dA$$

$$= e^{b^2(\beta\sigma)^2} \iint_{S_0} \sum_{j=1}^{N_c-1} R_j^m \left[\int_{-\infty}^{\infty} \frac{e^{-(\xi_j - b\beta\sigma^2)^2/2\sigma^2}}{\sqrt{2\pi} \, \sigma} \, d\xi_j \right.$$

$$\times \int_{\xi_j+(M_j-M_0)/b}^{\infty} \frac{e^{-(\xi_0 + b\beta\sigma^2)^2/2\sigma^2}}{\sqrt{2\pi} \, \sigma} \, d\xi_0$$

$$\times \prod_{\substack{i=1 \\ i \neq j}}^{N_c-1} \int_{\xi_j+(M_j-M_i)/b}^{\infty} \frac{e^{-\xi_i^2/2\sigma^2}}{\sqrt{2\pi} \, \sigma} \, d\xi_i \right] \kappa \, dA$$

$$= e^{b^2(\beta\sigma)^2} \iint_{S_0} \sum_{j=1}^{N_c-1} R_j^m \left[\int_{-\infty}^{\infty} \frac{e^{-z^2/2}}{\sqrt{2\pi}} \, Q\left(z + \frac{M_j - M_0}{b\sigma} + 2b\beta\sigma\right) \right.$$

$$\times \prod_{\substack{i=1 \\ i \neq j}}^{N_c-1} Q\left(z + \frac{M_j - M_i}{b\sigma} + b\beta\sigma\right) dz \right] \kappa \, dA. \tag{6.16}$$

Similarly, for the mean total interference to the zeroth base station from all users in the $\overline{S}_0$ region, we have

$$I_{\overline{S}_0} = \iint_{\overline{S}_0} \sum_{j=1}^{N_c} R_j^m \, E[10^{(\zeta_j - \zeta_0)/10}; \, r_j^m \, 10^{\zeta_j/10} < r_i^m \, 10^{\zeta_i/10} \quad \text{for all } i \neq j, \, i > 0] \, \kappa \, dA$$

$$= \iint_{\overline{S}_0} \sum_{j=1}^{N_c} R_j^m \, E[e^{\beta(\zeta_j - \zeta_0)}; \, \zeta_j + M_j < \zeta_i + M_i \quad \text{for all } i \neq j, \, i > 0] \, \kappa \, dA$$

$$= \iint_{\overline{S}_0} \sum_{j=1}^{N_c} R_j^m \, E[e^{\beta b(\xi_j - \xi_0)}; \, \xi_i > \xi_j + (M_j - M_i)/b \quad \text{for all } i \neq j, \, i > 0)] \, \kappa \, dA$$

$$\tag{6.17}$$

$$= e^{b^2(\beta\sigma)^2} \iint_{\overline{S}_0} \sum_{j=1}^{N_c} R_j^m \left[\int_{-\infty}^{\infty} \frac{e^{-(\xi_j - b\beta\sigma^2)^2/2\sigma^2}}{\sqrt{2\pi} \, \sigma} \, d\xi_j \right.$$

$$\times \prod_{\substack{i=1 \\ i \neq j}}^{N_c} \int_{\xi_j+(M_j-M_i)/b}^{\infty} \frac{e^{-\xi_i^2/2\sigma^2}}{\sqrt{2\pi} \, \sigma} \, d\xi_i \right] \kappa \, dA$$

$$= N_c \, e^{b^2(\beta\sigma)^2} \iint_{\overline{S}_0} R_1^m \left[\int_{-\infty}^{\infty} \frac{e^{-z^2/2}}{\sqrt{2\pi}} \prod_{i=2}^{N_c} Q\left(z + b\beta\sigma + \frac{M_1 - M_i}{b\sigma}\right) dz \right] \kappa \, dA.$$

Note that, in (6.17) the last step follows from the fact that subscripts 1 through N_c represent the N_c nearest cells and hence are interchangeable. Thus, the first subscripts can be made equal to 1 in each case without changing the result.

From these we obtain the relative interference at the zeroth cell from all users not controlled by its base station:

$$f = \frac{I_{S_0} + I_{\bar{S}_0}}{k_u}. \tag{6.18}$$

I_{S_0} and $I_{\bar{S}_0}$ are given by (6.16) and (6.17), respectively, with $\kappa = 2k_u/(3\sqrt{3})$. Also, as before, we may assume $b = 1/\sqrt{2}$. This general result specializes to the previous expressions (6.15) for $N_c = 2$ and (6.10) for $N_c = 1$. Table 6.3 contains the results of numerical integrations of (6.16) and (6.17) for $N_c = 3$ and 4. It also includes those for $N_c = 2$ and $N_c = 1$ of Tables 6.2 and 6.1, respectively, for $m = 4$ and several values of σ. Clearly, for $\sigma \le 6$ dB, use of the two nearest neighbors is enough to obtain most of the soft handoff advantage. For $\sigma = 8$ dB, use of three neighbors (to better deal with the cell's corners) provides a significant additional advantage.

Note finally that although we have implicitly assumed an omnidirectional base station antenna, because of the uniform user loading assumption that implies circular symmetry, all results apply also to a single sector of a multisectored cell antenna.

Table 6.3 Relative Other-Cell Interference Factor, f, for $N_c = 1$, 2, 3, and 4 ($m = 4$; $a^2 = \frac{1}{2}$)

Total Standard Deviation	$N_c = 1$	$N_c = 2$	$N_c = 3$	$N_c = 4$
0	0.44	0.44	0.44	0.44
2	0.48	0.43	0.43	0.43
4	0.67	0.47	0.45	0.45
6	1.13	0.56	0.49	0.49
8	2.38	0.77	0.57	0.55
10	6.17	1.28	0.75	0.66
12	19.8	2.62	1.17	0.91

6.5 Cell Coverage Issues with Hard and Soft Handoff

Before addressing capacity issues in the next section, we investigate the effect of soft handoff on cell coverage. We determine cell coverage for a virtually unloaded system, assuming a single user transmitting to either of two contiguous base stations in the presence of background noise only, with no other users present. We assume, as usual, log-normal shadowing and normalize the cell radius to unity.

6.5.1 Hard Handoff

Consider a hard handoff that occurs exactly at the boundary between cells (assuming an ideal condition established by an external observer). Then the relative attenuation from the mobile user at the boundary to either base station at distance r_i is given by (6.2),

$$10 \log \alpha(r_i, \zeta_i) \triangleq 10m \log r_i + \zeta_i, \qquad i = 1,2. \qquad (6.19)$$

Here, r_i are the normalized distances to the base stations, and ζ_i are the corresponding log-normal shadowing values with zero mean and standard deviation σ.

Without shadowing, the minimum power (measured in decibels) required from the mobile's transmitter just to overcome background noise is, of course, proportional to $10m \log r$. Normalizing the cell radius to $r = 1$, we eliminate this (common) term at the boundary. On the other hand, the random component due to shadowing, ζ, requires increased minimum power to guarantee the same performance most of the time. Suppose we require that the link achieve at least the performance of unshadowed propagation all but a fraction P_{out} of the time, which will be denoted the outage probability. A margin γ dB must then be added to the transmitted power. The desired performance will be achieved whenever the shadowing attenuation $\zeta < \gamma$. Thus the outage probability (or fraction of the time that the performance is not achieved) is

$$P_{out} = \Pr(\zeta > \gamma) = \frac{1}{\sqrt{2\pi}\,\sigma} \int_{\gamma}^{\infty} e^{-\zeta^2/2\sigma^2}\, d\zeta = Q\left(\frac{\gamma}{\sigma}\right). \qquad (6.20)$$

If we require that points on the boundary be adequately covered 90% of the time, then $P_{out} = 0.1$. For $\sigma = 8$ dB, it follows that the *margin* $\gamma = 10.3$ *dB*.

On the other hand, for hard handoff the ideal condition of handoff at the cell boundary is both unrealistic and undesirable. This is because it can lead to the "ping-pong" effect, where a user near the boundary is handed back and forth several times from one base station to the other. In practical hard-handoff systems, the handoff occurs only after the first cell's base station power is reduced far enough below its value at the boundary. Let us assume that this happens, with the required probability, when the user has moved a reasonable distance beyond the boundary. Let the ratio of this distance to the cell radius be $r' > 1$. At this distance from the first cell's base station, the outage probability, for a margin γ, becomes

$$P_{out} = \Pr(10m \log r' + \zeta > \gamma) = Q\left(\frac{\gamma - 10m \log r'}{\sigma}\right). \quad (6.21)$$

Therefore, for an outage probability $P_{out} = .1$ and $\sigma = 8$ dB, the required margin is

$$\gamma = 10m \log r' + 10.3 \text{ dB}. \quad (6.22)$$

The first two columns of Table 6.4 show the margin required for a propagation law $m = 4$ as a function of r', the relative distance beyond the cell boundary at which hard handoff occurs. Thus, for a relatively small range of additional distance beyond cell boundary for the hard handoff, the margin must be increased by 2 to 4 dB.

Table 6.4 Relative Margin and Coverage ($m = 4$; $a^2 = \frac{1}{2}$)

Relative Distance beyond Cell Boundary r'	Hard Handoff Required Margin γ_{Hard} (dB)	Relative Margin $\gamma_{Hard} - \gamma_{Soft}$ (dB)	Relative Coverage Area
1	10.3	4.1	1.6
1.05	11.1	4.9	1.8
1.1	12.0	5.8	2.0
1.15	12.7	6.5	2.1
1.2	13.5	7.3	2.3
1.25	14.2	8.0	2.5

6.5.2 Soft Handoff

For soft handoff, we now show that the required margin is much reduced. Soft handoff will occur throughout a range of distances from the two base stations. At any given time, or for any given frame or packet, the better of the two base stations' receptions will be used at the switching center. We assume for simplicity (and slightly pessimistically) that this depends only on attenuation. Then the lesser of the attenuations of (6.19) will apply. Thus, the margin γ needs only to satisfy the outage probability requirement,

$$P_{out} = \Pr\{Min[M_1 + \zeta_1, M_2 + \zeta_2] > \gamma\}, \tag{6.23}$$

where M_i is defined in (6.11).

But since ζ_1 and ζ_2 are correlated according to their definition (6.3), we may express them in terms of the independent variables ξ, ξ_1, and ξ_2, thus obtaining

$$
\begin{aligned}
P_{out} &= \Pr\{Min[M_1 + b\xi_1, M_2 + b\xi_2] > \gamma - a\xi\} \\
&= \frac{1}{(2\pi\sigma)^{3/2}} \int_{-\infty}^{\infty} e^{-\xi^2/2\sigma^2} \, d\xi \int_{(\gamma - a\xi - M_1)/b}^{\infty} e^{-\xi_1^2/2\sigma^2} \, d\xi_1 \\
&\quad \times \int_{(\gamma - a\xi - M_2)/b}^{\infty} e^{-\xi_2^2/2\sigma^2} \, d\xi_2 \\
&= \frac{1}{\sqrt{2\pi}} \int_{-\infty}^{\infty} e^{-x^2/2} \, Q\left(\frac{\gamma - M_1 - a\sigma x}{b\sigma}\right) Q\left(\frac{\gamma - M_2 - a\sigma x}{b\sigma}\right) dx.
\end{aligned}
\tag{6.24}
$$

Now since the mobile is assumed to be either in cell 1 or in cell 2, then either $r_1 \le 1, r_2 \ge 1$, or $r_1 \ge 1, r_2 \le 1$. It is easily established that $r_1 = r_2 = 1$ (where the mobile is exactly on the boundary) represents the worst case, so that $M_1 = M_2 = 0$ and

$$P_{out} \le \frac{1}{\sqrt{2\pi}} \int_{-\infty}^{\infty} e^{-x^2/2} \left[Q\left(\frac{\gamma - a\sigma x}{b\sigma}\right) \right]^2 dx. \tag{6.25}$$

For $a = b = 1/\sqrt{2}$ and $\sigma = 8$ dB, we find that for outage probability $P_{out} = 0.1$, the margin $\gamma_{soft} = 6.2$ dB.

Thus, as noted in the third column of Table 6.4, the required margin for soft handoff is (conservatively) about 6 dB to 8 dB less than for hard handoff. Cell area is proportional to the square of the radius while propagation loss is proportional to the fourth power. It follows that this margin reduction represents a cell area increase for soft handoff of 3 to 4 dB, or a reduction in the number of cells and consequently base stations by a factor[6] of 2 to 2.5, as shown in the last column of Table 6.4. This approach can be extended to the case where the cells are loaded as well [Vijayan *et al.*, 1994].[7] In this case, coverage will depend as well on the number of users per sector, and ultimately cell size will depend primarily on the number of users and their distribution in a particular geographical area. In the next section, we examine the ultimate cell capacity independent of size. There again we find that soft handoff, which reduces other-cell interference as shown in Section 6.4, also supports significantly greater capacity than hard handoff.

6.6 Erlang Capacity of Reverse Links

6.6.1 Erlang Capacity for Conventional Assigned-Slot Multiple Access

The capacity of a multiple access network is measured by the average number of users receiving service at a given time with a given level of quality, which includes requirements for both accuracy and service availability. For conventional frequency-division and time-division multiple access (which we shall call assigned-slot networks because a slot in time or frequency, or both, is assigned for each call), queuing analysis was first performed nearly a century ago for wired telephone traffic by Erlang [Bertsekas and Gallager, 1987]. This established capacity as a function of availability. Availability is defined as the complement of the probability that a user does not receive service at any given time because all slots are currently assigned to calls—a situation that evokes a busy signal. In a wireless system, the total number of available slots depends on total bandwidth, data rate per user, and frequency reuse factor, all of which determine the quality of the call in terms of availability or accuracy or both. We defer these issues to the end of this section. For nonslotted sys-

---

[6] Note, however, that this assumes the same spread spectrum system with hard or soft handoff. If a soft-handoff spread spectrum system is compared to a hard-handoff conventional slotted narrowband system, where both are operating in multipath fading, the difference may be considerably greater.

[7] From an economic viewpoint, coverage is the principal concern at the inception of service when the least number of base stations for a lightly loaded system dictates initial capital investment. As the system becomes loaded, capacity becomes the primary issue, with the number of base stations dictated primarily by the subscriber population.

tems employing spread spectrum multiple access, there are other variables besides traffic intensity, and these will be treated in 6.6.2.

Typically, for a large population of users, the call arrival rate per user is fairly small, but the total average arrival rate from the entire population, λ calls/s, may be large. Arrivals occur randomly at Poisson-distributed intervals. This is equivalent to modeling the arrival process [Bertsekas and Gallager, 1987] as a sequence of independent binary variables, in successive infinitesimal time intervals Δt, with a single arrival per interval occurring with probability $\lambda(\Delta t)$. The call service time per user is assumed to be exponentially distributed, so that the probability that service time τ exceeds T is given by

$$Pr(\tau > T) = e^{-\mu T}, \qquad T > 0. \tag{6.26}$$

From this it follows that the average call duration is $1/\mu$ s. Continuing with the infinitesimal independent increment model, this means that for a single server (slot), the probability that the call terminates during an interval of duration Δt seconds is

$$\begin{aligned} Pr(T < \tau < T + \Delta t \mid \tau > T) &= [e^{-\mu T} - e^{-\mu(T + \Delta t)}]/e^{-\mu T} \\ &= 1 - e^{-\mu \Delta t} \sim \mu(\Delta T) + o(\Delta t), \end{aligned} \tag{6.27}$$

where $o(\Delta t) \to 0$ as $\Delta t \to 0$. The number of time or frequency slots K_0 equals the maximum number of users that can be served simultaneously. Thus, as long as the number of active calls $k \leq K_0$, all k will be serviced. Then the probability that at least one will terminate in Δt is on the order of $k\mu(\Delta t)$. Note that the probability that more than one terminates is $o(\Delta t)$, which becomes negligible as $\Delta t \to 0$.

There are two commonly used models for determining the occupancy distribution and the probability of lost calls. These are illustrated by the Markov chains of Figures 6.3a and 6.3b, where the states represent the number of users in the system. They are called the "lost call cleared" (LCC) and "lost call held" (LCH) models. In the first (LCC), it is assumed that if a new user attempts to enter the network when all slots (servers) are occupied, it departs and reenters later after a random interval. In that case it is treated as a new user and thus slightly increases λ. The total number of states is therefore $1 + K_0$. In the second (LCH), it is assumed that the unserved users repeat their attempts to place a call immediately and thus remain in the system, though unserved. The result is that the number of states is infinite, and the average time per call $(1/\mu)$ is increased slightly so

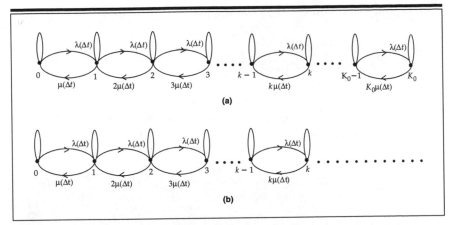

Figure 6.3 Markov state diagrams for Erlang traffic models. (a) LCC model. (b) LCH model. All self loop probabilities are such as to make outgoing branch probabilities sum to unity.

that μ is decreased slightly. The net effect for both models is to increase λ/μ slightly.

The Markov state equations for the LCC model of Figure 6.3a, in terms of the steady-state occupancy probabilities $P_0, P_1, \ldots, P_{K_0}$, are then

$$P_0 = [1 - \lambda(\Delta t)]P_0 + \mu(\Delta t)P_1, \tag{6.28}$$

$$P_k = \lambda(\Delta t)P_{k-1} + [1 - (\lambda + k\mu)\,\Delta t]P_k + (k+1)\mu(\Delta t)P_{k+1}, \\ k = 1, 2, \ldots, K_0 - 1, \tag{6.29}$$

$$P_{K_0} = \lambda(\Delta t)P_{K_0-1} + [1 - K_0\mu(\Delta t)]P_{K_0}. \tag{6.30}$$

It follows by induction on (6.29) and from (6.28) and (6.30) that

$$P_k = \frac{(\lambda/\mu)}{k}P_{k-1}, \quad k = 1, 2, \ldots, K_0. \tag{6.31}$$

Hence,

$$P_k = \frac{(\lambda/\mu)^k}{k!}P_0, \quad k = 1, 2, \ldots, K_0, \tag{6.32}$$

and

$$1 = \sum_{k=0}^{K_0} P_k = P_0 \sum_{k=0}^{K_0} \frac{(\lambda/\mu)^k}{k!}. \tag{6.33}$$

Combining (6.32) and (6.33) yields the occupancy distribution

$$P_k = \frac{(\lambda/\mu)^k/k!}{\sum\limits_{j=0}^{K_0} (\lambda/\mu)^j/j!}, \qquad k = 0, 1, \ldots, K_0 \qquad \text{(LCC)}, \quad (6.34)$$

as well as the probability that a new user enters the system and finds all servers busy, which is

$$P_B = P_{K_0} = \frac{(\lambda/\mu)^{K_0}/K_0!}{\sum\limits_{j=0}^{K_0} (\lambda/\mu)^j/j!} \qquad \text{(LCC)}. \qquad (6.35)$$

This last expression, called the *Erlang-B formula,* is the probability of lost calls or blocking probability in the LCC model.

For the LCH model of Figure 6.3b, (6.28) still holds, but now (6.29) holds for all integers k and (6.30) no longer applies (equivalently, we may let $K_0 \to \infty$). Thus, (6.31) and (6.32) now hold for all k, and hence

$$1 = \sum_{k=0}^{\infty} P_k = P_0 \sum_{k=0}^{\infty} \frac{(\lambda/\mu)^k}{k!} = P_0 e^{\lambda/\mu}.$$

Thus, for the LCH model,

$$P_k = \frac{(\lambda/\mu)^k}{k!} e^{-\lambda/\mu}, \qquad k = 0, 1, 2, \ldots \qquad \text{(LCH)}, \quad (6.36)$$

which is the Poisson formula. The blocking probability here is the probability that when a new user arrives there are K_0 *or more users* in the system, either being served or seeking service. Hence, the *blocking probability* for the LCH model is

$$P_B = \sum_{k=K_0}^{\infty} P_k = e^{-\lambda/\mu} \sum_{k=K_0}^{\infty} (\lambda/\mu)^k/k! \qquad \text{(LCH)}, \qquad (6.37)$$

which is just the sum of the tail of a Poisson distribution.

We shall favor the latter (LCH) model both because it is more realistic for mobile users and because it more closely resembles the case for un-slotted multiple access. As already noted, for large K_0 the results obtained for the two models are very similar. In both cases λ/μ is slightly increased by lost calls reentering, with λ increased in the LCC model and μ decreased in the LCH model.

All that remains is to relate the number of slots K_0 to W, R, and the frequency reuse factor, Φ. This, of course, depends on the modulation (and coding) efficiency. If a modulation is used that can support n bits/s/Hz (which refers to bits prior to coding expansion), then for a sectored cell, the number of slots per sector is

$$K_0 = (W/R)n\Phi \text{ slots/sector.} \tag{6.38}$$

However, n is related to Φ because the larger the reuse factor, the more interference arrives from other cells, and hence the lower the modulation efficiency n. For two conventional systems, employing TDMA standards adopted in Europe and North America [Steele, 1992], the parameters used in a three-sectored cell are $n \approx 0.5$, $\Phi = 1/12$, and $n \approx 1.0$, $\Phi = 1/21$, respectively. Thus, for these two TDMA systems $K_0 \approx (W/R)/24$ slots/sector and $K_0 \approx (W/R)/21$ slots/sector, respectively.

6.6.2 Spread Spectrum Multiple Access Outage— Single Cell and Perfect Power Control

In spread spectrum systems, since all users occupy the same frequency spectrum and time allocation, there are no slots. As has been noted previously, the system is strictly interference limited. Following the development of Viterbi and Viterbi (1993), throughout the rest of Section 6.6, we assume initially a single cell occupied by k_u perfectly power-controlled users, so that each is received by the base station at the same power level. However, each user's digital input may be intermittent: a voice call with variable rate based on voice activity detection, or an interactive message that requires a response before sending new data. During inactive periods, the user's signal power is suppressed. Then for a total bandwidth occupancy W Hz for all users with equal data rates R bits/s, background noise N_0 watts/Hz and equal, perfectly controlled bit energies E_b, the total average power received by the cell, assuming stationary arrivals and user activity, is

$$\text{Total power} = \sum_{i=1}^{k_u} \nu_i \, E_b R + N_0 W. \tag{6.39}$$

Here, ν_i is a binary random variable indicating whether or not the ith user is active at any instant, and

$$\rho \triangleq \Pr(\nu_i = 1) = 1 - \Pr(\nu_i = 0) \tag{6.40}$$

is the "activity factor" or fraction of the time during which the user's signal is present. Since the total received power is the sum of noise, interference power, and the desired user power (which we denote by subscript 1), we have from (6.39), that the average noise-plus-interference power, denoted $I_0 W$, is

$$I_0 W = \sum_{i=2}^{k_u} \nu_i E_b R + N_0 W. \tag{6.41}$$

For dynamic range limitations on the multiple access receiver of bandwidth W (as well as to guarantee system stability, as will become clear later), it is desirable to limit the total received noise-plus-interference power-to-background noise, or equivalently the ratio $I_0 W / N_0 W$. Thus, we require

$$\frac{I_0}{N_0} < \frac{1}{\eta}, \qquad \text{where } \eta < 1, \tag{6.42}$$

and typically $\eta = .25$ to $.1$, corresponding to power ratios $I_0/N_0 = 6$ dB to 10 dB. Combining this condition with (6.41) yields the requirement

$$\sum_{i=2}^{k_u} \nu_i E_b R = (I_0 - N_0)W < I_0 W(1 - \eta)$$

or

$$\sum_{i=2}^{k_u} \nu_i < \frac{(W/R)(1 - \eta)}{E_b/I_0} \triangleq K_o(1 - \eta) \triangleq K_o', \tag{6.43}$$

where both the ν_i and k_u are independent random variables. *When condition (6.43) is not met,* the system will be deemed to be in the *outage condition.* This condition is only temporary, remedied without intervention by the random variation of the variables. The power control mechanism can accelerate the process and guarantee stability by reducing the E_b/I_0 requirement for all users. In either case, the probability of outage, P_{out} (which is similar to, but more benign than, the blocking probability for the slotted systems treated in the previous subsection), is upper-bounded slightly by including the desired signal variable, v_1 in the summation. Therefore,

$$P_{out} = \Pr\left(\sum_{i=2}^{k_u} v_i > K'_o\right) < \Pr\left(\sum_{i=1}^{k_u} v_i > K'_o\right). \tag{6.44}$$

Thus, the outage probability is determined from the distribution of the sum of k_u independent binary variables, each with probability given by (6.40). But k_u, the number of users in the network, is also a random variable. In fact, the user arrival and service model described in Section 6.6.1 for slotted systems applies equally here. Since users remain in the system through outage, the model for LCH (Figure 6.3b) applies. Thus, the distribution of k_u is given by (6.36), and the distribution of the random variable of interest,

$$Z \triangleq \sum_{i=1}^{k_u} v_i, \tag{6.45}$$

can best be determined from its moment-generating function,

$$
\begin{aligned}
E(e^{sZ}) &= E_{k_u} \prod_{i=1}^{k_u} E_{v_i}(e^{sv_i}) \\
&= E_{k_u}[\rho e^s + (1 - \rho)]^{k_u} \\
&= \sum_{k_u=0}^{\infty} \frac{(\lambda/\mu)^{k_u}}{k_u!} e^{-\lambda/\mu} [\rho(e^s - 1) + 1]^{k_u} = \exp\left[\frac{\lambda\rho}{\mu}(e^s - 1)\right],
\end{aligned}
\tag{6.46}
$$

where we have used (6.40) and (6.36). This is the generating function of a Poisson distribution. We can best see this by noting that if $\rho = 1$, according to (6.45), $Z = k_u$, which is Poisson. Thus, the binary random variables do not increase the "randomness" of the Poisson process; they merely reduce the occupancy rate (λ/μ) by the factor ρ. Thus, the outage probability of (6.44) is just the sum of the Poisson tail,

$$P_{out} < e^{-\rho\lambda/\mu} \sum_{k=\lfloor K'_o \rfloor}^{\infty} (\rho\lambda/\mu)^k/k!. \tag{6.47}$$

The difference between this and (6.37) is that, as just noted, the λ/μ parameter is decreased by $\rho < 1$. Also, while in the slotted case K_0 was the effective number of slots per cell (reduced by the reuse factor), here $K_0 =$

$(W/R)/(E_b/I_0)$, and K_0' is further reduced by the backoff factor $(1 - \eta)$, as given by (6.42) and (6.43).

Although the tail of the Poisson distribution can be computed numerically, from the moment-generating function we can readily obtain the Chernoff bound:

$$P_{out} < \operatorname*{Min}_{s>0} E\{\exp[s(Z - K_0')]\}$$

$$= \operatorname*{Min}_{s>0} \exp\left[\frac{\lambda\rho}{\mu}(e^s - 1) - sK_0'\right] \qquad (6.48)$$

$$= \exp\left\{-K_0'\left[\ln\left(\frac{K_0'}{\rho\lambda/\mu}\right) - 1 + \frac{\rho\lambda/\mu}{K_0'}\right]\right\}.$$

Another approximation, accurate for large K_0' [Feller, 1957], is obtained through approximating the Poisson by a Gaussian variable with the same mean and variance, both being $\rho\lambda/\mu$. This yields

$$P_{out} \approx Q\left(\frac{K_0' - \rho\lambda/\mu}{\sqrt{\rho\lambda/\mu}}\right). \qquad (6.49)$$

Both (6.48) and (6.49) are shown in Figure 6.4 as a function of $\rho\lambda/\mu$ for $K_0 = 256$ and $\eta = .1$ ($K_0' = 230.4$). Note that these apply equally to the

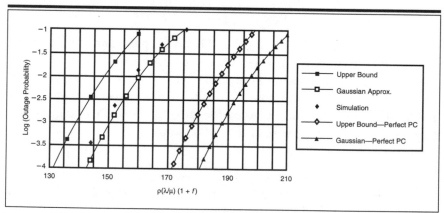

Figure 6.4 Reverse link outage probabilities. $K_0 = 256$, $\eta = 0.1$, $I_0/N_0 = 10$ dB, power control $\sigma_C = 2.5$ dB ($f = 0$ for single cell).

slotted case if we set $\rho = 1$ and $K_o' = K_o$, the effective number of slots per cell, as given by (6.38).[8]

6.6.3 Outage with Multiple-Cell Interference

We determined in Section 6.4 that, assuming uniform loading of all cells, users controlled by other cell base stations introduce interference power into a given base station whose average is $f k_u$, where k_u is the average number of active users per cell. With soft handoff, f lies between 0.5 and 0.6 for a propagation power law $m = 4$ and a log-normal shadowing standard deviation $\sigma \approx 8$ dB. Furthermore, because of the nature of soft handoff, no other-cell user can arrive at the given base station with power higher than the same-cell users, for otherwise it would be handed off. Continuing with the perfect power control assumption, which we shall abandon in the next subsection, this means that all other-cell users can be accounted for in the outage probability expression (6.44) by additional terms having $\nu_i \leq 1$, when active.

Since the total contribution from these additional terms has a mean value of $f k_u$, instead of determining the complete distribution of each term, we model their effect as that of this many additional users with maximum normalized power value, unity. Thus, (6.44) becomes

$$P_{out} = \Pr \left[\sum_{i=1}^{k_u(1+f)} \nu_i > K_o' \right]. \tag{6.50}$$

This model appears pessimistic because it creates a more coarsely quantized distribution, whereas in reality the random variable sum in (6.50) is a continuous variable. Note that these additional users also transmit intermittently with $\Pr(\nu = 1) = \rho$ and all cells have the same Poisson occupancy distribution as those in the given cell. Thus, $k_u' \triangleq k_u(1+f)$ is also a Poisson variable with parameter $\rho(\lambda/\mu)(1+f)$. With this model, then, the outage probability Chernoff bound of (6.48) and normal approximation of (6.49) hold as before, with $\rho\lambda/\mu$ increased by the factor $(1+f)$, as shown in Figure 6.4.

---

[8] It may seem that the main capacity advantage of CDMA over slotted multiple access is due to the voice (or message) activity factor $\rho < 1$. Rather, as will be shown in Section 6.8, an equal or greater advantage is derived from the sectorized antenna gain [since Φ is inversely proportional to the number of sectors, while (6.49) is independent thereof], as well as the mitigation of other-cell interference to be treated next.

6.6.4 Outage with Imperfect Power Control

We come finally to the most important issue of all, the effect of imperfect power control on capacity. Based on the analytical treatment in Section 4.7 and the discussion in Section 6.2, as well as experimental measurements [Padovani, 1994; Viterbi and Padovani, 1992], a user that is controlled to a desired E_b/I_0 level (which level may be varied according to multipath propagation conditions) will exhibit a received E_b/I_0 level at the desired cell base station that varies according to a log-normal distribution with a standard deviation on the order of 1.5 to 2.5 dB. With this condition, the outage probability derivation of (6.39) through (6.49) needs to be modified so that the constant value of E_b for all users is replaced by a variable $E_{b_i} \triangleq \epsilon_i E_{b_0}$, which is log-normally distributed. (We shall determine the appropriate value of scale factor E_{b_0} in what follows.) By so doing, and accounting also for other-cell interference, we obtain in place of (6.44) and (6.45)

$$P_{out} < \Pr\left[Z' = \sum_{i=1}^{k'_u} \epsilon_i \nu_i > K'_o \right]. \tag{6.51}$$

$k'_u = k_u (1 + f)$ is a Poisson variable with parameter $(\lambda/\mu)(1 + f)$, while $E_{b_0} \epsilon_i$ is log-normally distributed. Thus, we may define

$$x_i = 10 \log_{10}(\epsilon_i E_{b_0}/I_0) \tag{6.52}$$

to be a normally distributed random variable of mean m_c and standard deviation σ_c. Hence, inverting (6.52), we have

$$(E_{b_0}/I_0)\, \epsilon_i = 10^{x_i/10} = e^{\beta x_i}, \qquad \text{where } \beta = \ln(10)/10. \tag{6.53}$$

Before we try to find the moment-generating function, we evaluate the nth moment of ϵ_i using the fact that x_i is Gaussian with mean m_c and standard deviation σ_c:

$$
\begin{aligned}
E(\epsilon_i^n) &= (E_{b_0}/I_0)^{-n} \int_{-\infty}^{\infty} e^{n\beta x_i} \frac{e^{-(x_i - m_c)^2/2\sigma_c^2}}{\sqrt{2\pi\sigma_c^2}}\, dx_i \\
&= \frac{(e^{\beta m_c})^n}{(E_{b_0}/I_0)^n} e^{n^2(\beta\sigma_c)^2/2}.
\end{aligned}
\tag{6.54}
$$

Since E_{b_0} was defined as an arbitrary constant, it makes sense to choose it such that

$$E_{b_0}/I_0 = e^{\beta m_c} = 10^{m_c/10},\qquad(6.55)$$

which equals the numerical equivalent of the decibel mean. More important, it can be seen that since $x_i < m_c$ with probability $\frac{1}{2}$, $(E_b/I_0)_i = (E_{b_0}/I_0)\,\epsilon_i < e^{\beta m_c}$ with probability $\frac{1}{2}$, meaning that

$$E_{b_0}/I_0 = \text{median}\ [(E_b/I_0)_i]\qquad\text{for all }i.\qquad(6.56)$$

Using (6.55) in (6.54), the nth moment of ϵ_i becomes

$$E(\epsilon_i^n) = e^{n^2(\beta\sigma_c)^2/2}.\qquad(6.57)$$

Unfortunately, the moment-generating function of ϵ does not exist, because its moment series expansion does not converge. However, a modified Chernoff bound for outage probability can still be obtained by using a truncated moment-generating function approach. Suppose we break up the outage probability expression (6.51) into two parts, where the first is conditioned on $\nu_i\,\epsilon_i < T$ for all i for some sufficiently large T, and the second is conditioned on the complementary event. Thus, upper-bounding the expression by unity for the second part,

$$
\begin{aligned}
P_{\text{out}} &< \Pr\left[\sum_{i=1}^{k'_u}\nu_i\,\epsilon_i > K'_0;\ \nu_i\,\epsilon_i < T \text{ for all } i\right]\\
&\quad + \Pr[\nu_i\,\epsilon_i > T \text{ for any } i]\\[4pt]
&< E_{k'_u}\sum_{i=1}^{k'_u} E_{\nu_i}(e^{s\nu_i\epsilon_i};\ \epsilon_i < T)e^{-sK'_0}\\
&\quad + E_{k'_u}\left[\sum_{i=1}^{k'_u}\Pr(\epsilon_i > T)\Pr(\nu_i = 1)\right]\\[4pt]
&< \underset{s>0,T>0}{\text{Min}}\ \{E_{k'_u}\{[\rho E_{\epsilon}e^{s\epsilon} + (1-\rho)]^{k'_u};\ \epsilon < T\}\\
&\qquad\times e^{-sK'_0} + E_{k'_u}[k'_u\Pr(\epsilon > T)\,\rho]\}\\[4pt]
&= \underset{s>0,T>0}{\text{Min}}\ \{\exp[\rho(\lambda/\mu)(1+f)\,E(e^{s\epsilon_T}-1) - sK'_0]\\
&\qquad + \rho(\lambda/\mu)(1+f)\,\Pr(\epsilon > T)\}.\qquad(6.58)
\end{aligned}
$$

Here,

$$E(e^{s\epsilon_T}) = E[\exp(se^{\beta\xi}); \xi \le \ln T/\beta]$$

and

$$\Pr(\epsilon > T) = \Pr(\xi > \ln T/\beta),$$

and $\xi \triangleq (\ln \epsilon)/\beta = x - m_c$, according to (6.53) through (6.55). Thus, ξ is Gaussian with zero mean and variance σ_c^2, and, defining $\theta \triangleq (\ln T)/\beta$,

$$\Pr(\epsilon > T) = \frac{1}{\sqrt{2\pi}\sigma_c} \int_{(\ln T)/\beta}^{\infty} e^{-\xi^2/2\sigma_c^2} \, d\xi = Q\left(\frac{\ln T}{\beta\sigma_c}\right) = Q\left(\frac{\theta}{\sigma_c}\right), \quad (6.59)$$

while

$$E(e^{s\epsilon_T}) = \frac{1}{\sqrt{2\pi}\,\sigma_c} \int_{-\infty}^{\theta} e^{se^{\beta\xi}} e^{-\xi^2/2\sigma_c^2} \, d\xi. \quad (6.60)$$

The truncated moment-generating function (6.60) can be computed numerically or from the series expansion

$$\begin{aligned}
E(e^{s\epsilon_T}) &= \sum_{n=0}^{\infty} \frac{s^n}{n!\sqrt{2\pi}\,\sigma_c} \int_{-\infty}^{\theta} e^{n\beta\xi} e^{-\xi^2/2\sigma_c^2} \, d\xi \\
&= \sum_{n=0}^{\infty} \frac{s^n}{n!} e^{n^2(\beta\sigma_c)^2/2} Q(n\beta\sigma_c - \theta/\sigma_c).
\end{aligned} \quad (6.61)$$

This series can alternatively be used for computing the truncated moment-generating function, since $Q(x) < e^{-x^2/2}$ for $x > 0$, so that

$$e^{n^2(\beta\sigma_c)^2/2} Q(n\beta\sigma_c - \theta/\sigma_c) < e^{-\theta^2/2\sigma_c^2} e^{n\beta\theta} \qquad \text{for } n > \theta/(\beta\sigma_c^2),$$

which guarantees convergence of (6.61).

Consequently, the outage probability with imperfect power control is bounded by (6.58) along with (6.59) and either (6.60) or (6.61). The bound is shown in Figure 6.4 as a function of $\rho(\lambda/\mu)(1 + f)$ for $K_o = 256$, $\eta = .1$, and $\sigma_c = 2.5$ dB.

As we did for perfect power control in Section 6.6.3, we can also obtain the Gaussian approximation. For this we need not truncate the moments, since the untruncated first and second moments exist, as given by (6.54).

Thus, approximating the distribution of Z' in (6.51), we obtain

$$P_{\text{out}} \approx Q\left[\frac{K'_0 - E(Z')}{\sqrt{\text{Var}(Z')}}\right].$$

But

$$E(Z') = E\left(\sum_{i=1}^{k'_u} \nu_i \, \epsilon_i\right) = E(k'_u)E(\nu)E(\epsilon) = (\lambda/\mu)(1 + f)\,\rho\,e^{(\beta\sigma_c)^2/2}, \quad (6.62)$$

while [Feller, 1957]

$$\text{Var}(Z') = \text{Var}\left(\sum_{i=1}^{k'_u} \nu_i \, \epsilon_i\right) = E(k'_u)\text{Var}(\nu\,\epsilon) + \text{Var}(k'_u)[E(\nu\,\epsilon)]^2.$$

But since k'_u is Poisson, its mean and variance are both equal to (λ/μ) $(1 + f)$, so that

$$\begin{aligned}
\text{Var}(Z') = \text{Var}\left(\sum_{i=1}^{k'_u} \nu_i \, \epsilon_i\right) &= (\lambda/\mu)(1 + f)\,E[(\nu\,\epsilon)^2] \\
&= (\lambda/\mu)(1 + f)\,\rho\,e^{2(\beta\sigma_c)^2}.
\end{aligned} \quad (6.63)$$

Thus, the normal approximation can be written

$$P_{\text{out}} \approx Q\left[\frac{K'_0 - \rho(\lambda/\mu)(1 + f)\,e^{(\beta\sigma_c)^2/2}}{\sqrt{\rho(\lambda/\mu)(1 + f)}\,e^{(\beta\sigma_c)^2}}\right]. \quad (6.64)$$

This approximation is also shown in Figure 6.4. While the truncated Chernoff bound of (6.58) is always valid, the Gaussian central limit theorem approximation is not as supportable as for the constant equal power case. It does appear reasonable, nevertheless, since Chernoff bounds usually overestimate probability by an order of magnitude. Yet to further substantiate the approximation, a simulation was performed using approximately one million events for each user. This result also appears in Figure 6.4. It agrees to within about 1% of the Gaussian approximation. For analytic simplicity we shall use the latter henceforth.

6.6.5 An Approximate Explicit Formula for Capacity with Imperfect Power Control

We may invert the approximate expression for outage probability (6.64) by solving a quadratic equation, to obtain the explicit formula for normalized average user occupancy, λ/μ in terms of *erlangs/sector*:

$$(\lambda/\mu)\rho(1 + f) = K_o' F(B, \sigma_c), \qquad (6.65)$$

where

$$B = \frac{[Q^{-1}(P_{out})]^2}{K_o'}, \qquad K_o' = \frac{W/R}{E_{b_0}/I_0}(1 - \eta), \qquad (6.66)$$

and

$$F(B, \sigma_c) = \frac{1}{\alpha_c}\left[1 + \frac{\alpha_c^3 B}{2}\left(1 - \sqrt{1 + \frac{4}{\alpha_c^3 B}}\right)\right]; \qquad \alpha_c = e^{(\beta\sigma_c)^2/2}. \qquad (6.67)$$

For convenience, plots of $(Q^{-1})^2$ and $F(B, \sigma_c)$ are given in Figures 6.5 and 6.6. The latter shows the degradation due to traffic variability (trunking efficiency) and power control. Note that setting $(1 - \eta) F(B, \sigma_c) = 1$ in

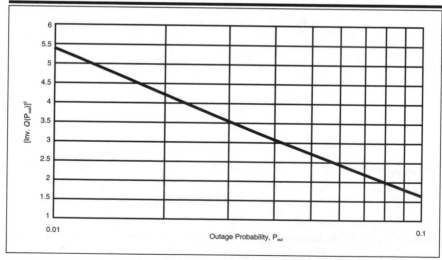

Figure 6.5 Variable factor in *B* as a function of outage probability. ©1993 IEEE. ``Erlang Capacity of a Power Controlled CDMA System'' by A. M. Viterbi and A. J. Viterbi, in *IEEE Journal on Selected Areas in Communications*, Vol. 11, No. 6, pp. 892–900, August 1993.

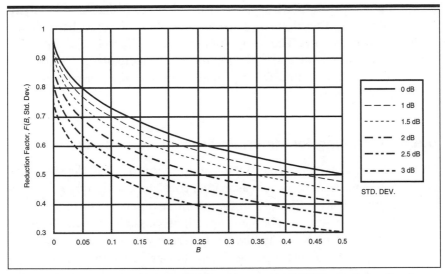

Figure 6.6 Erlang capacity reduction factors. ©1993 IEEE. "Erlang Capacity of a Power Controlled CDMA System" by A. M. Viterbi and A. J. Viterbi, in *IEEE Journal on Selected Areas in Communications,* Vol. 11, No. 6, pp. 892–900, August 1993.

(6.65), with proper identification of parameters, we obtain the naive formula of (1.5).

From this or, more accurately, from a comparison of both the two bounds and the two approximations for perfect and imperfect power control in Figure 6.4, it appears that imperfect power control with standard deviation, σ_c, as large as 2.5 dB results in a capacity reduction of no more than 1.0 dB or 20%.

We note that the formulas (6.65) through (6.67) apply also to slotted systems with K_0' replaced by K_0 of (6.38) and with $\sigma_c = 0, f = 0$, and $\rho = 1$. Hence, the ratio of Erlang capacities for spread spectrum and slotted multiple access systems, each in terms of erlangs/sector, is approximately

$$\frac{(\lambda/\mu)_{\text{spread}}}{(\lambda/\mu)_{\text{slotted}}} \approx \left\{ \frac{F(B_{\text{spread}}, \sigma_c)(1 - \eta)/[\rho(1 + f)]}{F(B_{\text{slotted}}, 0)} \right\} \left\{ \frac{1/(E_{b_0}/I_0)}{n\Phi} \right\}.$$

With parameter values $\rho = 0.4, f = 0.6, \eta = .1$, and $E_{b_0}/I_0 = 6$ dB, we can verify using (6.66) and (6.67) or Figures 6.5 and 6.6 that for a wide range of W/R, the first term exceeds unity. Setting it equal to one to obtain a simpler though cruder approximation, we conclude that roughly,

$$\frac{(\lambda/\mu)_{\text{spread}}}{(\lambda/\mu)_{\text{slotted}}} \approx \frac{1/(E_{b_0}/I_0)}{n\Phi}.$$

To compare a spread CDMA system with the two TDMA slotted systems cited at the end of Section 6.6.1, we assume three-sectored base station antennas in each case. Then with parameter $E_{b_o}/I_0 = 4$ (6 dB) for the former and $n\Phi = 1/24$ and $1/21$ for the latter, as explained in Section 6.6.1, the spread-to-slotted Erlang capacity ratios are approximately 6:1 and 5:1, respectively.

6.6.6 Designing for Minimum Transmitted Power

Until now, we have selected the interference-to-noise density ratio $I_0/N_0 = 1/\eta$ to be as large as reasonable, consistent with dynamic range and stability considerations, so as to maximize capacity. However, when the traffic occupancy is relatively low, it is possible instead to select smaller values of I_0/N_0, or equivalently larger values of $\eta < 1$, so as to bring down the transmitted power for each user. Thus, for any $\eta < 1$ we may use (6.65) through (6.67) to determine λ/μ required to establish the desired outage probability P_{out}. From this value of $\eta = \eta^*$ and the corresponding value of $\lambda/\mu = (\lambda/\mu)^*$, we may then obtain the required average user signal-to-background noise ratio at the base station from the expression for total average received signal-to-noise at the base station (for all the users of that cell). This expression is

$$\rho\left(\frac{\lambda}{\mu}\right)^* \frac{P_s}{N_o W} = \frac{I_0^* - N_0}{N_0} = \frac{1}{\eta^*} - 1.$$

Thus, the per-user received power (when the user is active) is

$$\frac{P_s}{N_0 W} = \frac{1}{\rho(\lambda/\mu)^*}\left(\frac{1}{\eta^*} - 1\right). \tag{6.68}$$

This is plotted in Figure 6.7 for $P_{out} = .01$, $K_o = 256$, $\sigma_c = 2.5$ dB, and $f = 0.55$. Capacity is normalized by that achieved with $\eta^* = 0.1$. Note that the required average received power, and hence also the transmitted power, per user is decreased by about 8 dB as the network loading λ/μ decreases by an order of magnitude. In other words, when the network is lightly loaded, a user at a given location needs to expand only about one-seventh as much transmitted power to achieve the desired performance as when the network is loaded to capacity.

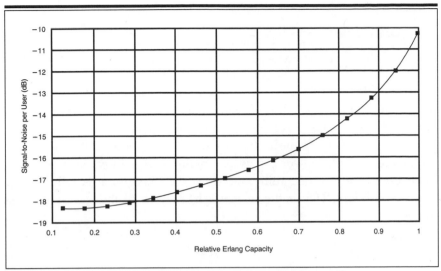

Figure 6.7 Signal-to-noise per user as a function of relative sector capacity. ©1993 IEEE. ``Erlang Capacity of a Power Controlled CDMA System'' by A. M. Viterbi and A. J. Viterbi, in *IEEE Journal on Selected Areas in Communications,* Vol. 11, No. 6, pp. 892–900, August 1993.

6.6.7 Capacity Requirements for Initial Accesses

In conventional slotted networks, certain slots are set aside for the multiple access users to request a call or channel allocation by the base station. A protocol is provided for recovering from collisions that occur when two or more new users send access requests simultaneously [Bertsekas and Gallager, 1987]. With spread spectrum multiple access, the allocated commodity is energy, rather than time or frequency. Thus, access requests can share the common channel with ongoing users. The arrival rate of user requests equals the cell arrival rate, λ calls/second, under the assumption that all requests are eventually served. Newly arriving users are not power-controlled until their requests are recognized, so the initial power level is a random variable uniformly distributed from 0 to a maximum value, corresponding to a received energy (per bit) level of E_M. Thus, the initial access energy level is γE_M, where γ is a random variable with probability density function

$$p(\gamma) = \begin{cases} 1, & 0 < \gamma < 1, \\ 0, & \text{otherwise,} \end{cases}$$

and distribution function

$$Pr(\gamma < x) = \begin{cases} 0, & x < 0, \\ x, & 0 \le x \le 1, \\ 1, & x > 1. \end{cases} \tag{6.69}$$

If this initial power level is not sufficient for detection,[9] and hence acknowledgment is not received, the user increases its power in constant decibel steps every frame until its request is acknowledged. Thus, the initial access user's power grows exponentially (i.e., linearly in decibels), with time taken to be continuous, since the frame time is only tens of milliseconds long. Hence, the energy as a function of time for initial access requests is

$$E(t) = \gamma E_M e^{\delta t}, \tag{6.70}$$

where δ is fixed. Letting τ_A denote the time required for an initial access user to be detected, it follows that

$$Pr(\tau_A > T_A) = Pr(\gamma E_M e^{\delta T_A} < E_M) = Pr(\gamma < e^{-\delta T_A})$$
$$= e^{-\delta T_A}, \quad T_A > 0, \tag{6.71}$$

where the last equation follows from (6.69).

Thus, the "service time" (time required for acceptance) for initial access requests for any user is exponentially distributed with mean $1/\delta$ s/message. The consequences of this observation are significant: With Poisson distributed arrivals and exponentially distributed service times, according to Burke's theorem [Bertsekas and Gallager, 1987] *the output distribution is also Poisson.* This guarantees that the arrival distribution for users initiating service is the same as for users who are just starting to seek access. Furthermore, at any given time the power distribution for users that have not yet been accepted is also uniform.

Thus, the total interference is augmented by the interference from initial accesses. When normalized, this introduces an additional term in the

---

[9] We assume, initially, that the access request is detected (immediately) when the initial access user's energy reaches E_M, but it remains undetected until that point. Below, we shall take E_M to be itself a random variable, with distribution similar to that of E_b. It should also be noted that initial access detection is performed on an unmodulated signal, other than for the pseudorandom code, which is known to the base station. It should thus be more robust than the demodulator performance at the same power level.

mean and variance of Z', (6.62) and (6.63), respectively. Letting this augmented interference variable be Z'', assuming that surrounding cells have the same arrival rate, we have

$$E(Z'') = E(Z') + (\lambda/\delta)(1 + f)\, E(\gamma E_M/I_0)/(E_{b_0}/I_0), \qquad (6.72)$$

$$\mathrm{Var}(Z'') = \mathrm{Var}(Z') + (\lambda/\delta)(1 + f)\, E[(\gamma E_M/I_0)^2]/(E_{b_0}/I_0)^2, \quad (6.73)$$

where γ is uniformly distributed on the unit interval and is independent of E_M/I_0. Thus,

$$E(\gamma) = \tfrac{1}{2} \qquad \text{and } E(\gamma^2) = \tfrac{1}{3}. \qquad (6.74)$$

Further, if we assume for a fixed θ, $E_M/I_0 = \theta(E_b/I_0)$, where E_b/I_0 is the required value for the given user after service has been provided,

$$E(E_M/I_0) = \theta(E_{b_0}/I_0)\, E(\epsilon), \qquad E[(E_M/I_0)^2] = \theta^2(E_{b_0}/I_0)^2\, E(\epsilon^2). \quad (6.75)$$

Thus, from (6.62), (6.63), and (6.72) through (6.75), we obtain

$$E(Z'') = \{(\lambda/\mu)\rho \exp[(\beta\sigma_c)^2/2] + (\lambda/\delta)(1/2)\theta \exp[(\beta\sigma_c)^2/2]\} \\ \times (1 + f) \qquad (6.76)$$

$$= \rho(\lambda/\mu)\left(1 + \frac{\theta\mu}{2\rho\delta}\right)\exp[(\beta\sigma_c)^2/2](1 + f),$$

$$\mathrm{Var}(Z'') = \{(\lambda/\mu)\rho \exp[2(\beta\sigma_c)^2] + (\lambda/\delta)(1/3)\theta^2 \exp[2(\beta\sigma_c)^2]\} \\ \times (1 + f) \qquad (6.77)$$

$$= \rho(\lambda/\mu)\left(1 + \frac{\theta^2\mu}{3\rho\delta}\right)\exp[2(\beta\sigma_c)^2](1 + f).$$

It is reasonable to take the arbitrary scale factor $\theta = \tfrac{3}{2}$, because this places the detection bit energy level at 50% higher than the operating bit energy level. We find that the effect is to increase the Erlang level (λ/μ) by the factor

$$1 + \left(\frac{3}{4\rho}\right)\left(\frac{\mu}{\delta}\right) = 1 + 1.88(\mu/\delta) \qquad \text{for } \rho = .4.$$

Note that μ/δ is the ratio of mean detection time to mean message duration, which should be very small. For example, if $\mu/\delta = .0055$, corresponding to a mean access time of one second for a three-minute mean call duration, then initial accesses reduce the supportable Erlang traffic by about 1%. This cost is negligible, considering the notable advantage of sharing the channel for initial access and ongoing traffic.

6.7 Erlang Capacity of Forward Links

6.7.1 Forward Link Power Allocation

As was pointed out from the outset, the forward link implementation and performance is vastly different from that of the reverse link. This is due primarily to three factors:

(a) access is one-to-many instead of many-to-one;

(b) synchronization and coherent detection are facilitated by use of a common pilot signal;

(c) the interference is received from a few concentrated large sources (base stations) rather than many distributed small ones (mobiles).

At the same time, the dynamic range of required power level will be much less for the forward link, since interference from same-cell users does not depend on distance from the base station. The worst-case situation for a mobile receiver will generally occur when the mobile is in one of the six corners of the hexagonal cell, where it is equidistant from three base stations.

More forward link power must then be provided to users that receive the most interference from other base stations. Of course, users on the boundaries may be in soft handoff, in which case they also receive signal power from two or more base stations. We defer consideration of soft handoff until the next subsection and consider here only reception of the desired user signal from a single base station. We assume that all base stations transmit at the same total average power level, but that the different users of a given cell are allocated different power levels according to their relative needs. We shall refer to this procedure as forward power *allocation*, rather than power control: The allocation is of a resource shared by many users, and it varies slowly and over a much smaller dynamic range. We follow mostly the development of Gilhousen *et al.* [1991].

Suppose that user i, controlled by base station 1, receives interference power from J base stations. Let the total power from base station j received by user i be $S_{R_{ji}}$, and let $S_{R_{1i}} > S_{R_{ji}}$ for all $j \neq 1$, for the user should always communicate with (at least) the base station from which it receives the strongest signal. Now suppose that a fraction $1-\beta$ of the total power transmitted by any cell or sector is devoted to the pilot signal, as well as to any common information (such as the initial vector of the pseudorandom code or paging data) destined to all users. The remaining fraction β is then allocated to all k_u users of the cell's base station or of a sector thereof. Thus, the ith user receives a fraction $\beta\phi_i$ of the transmitted power, under the conditions

$$\sum_{i=1}^{k_u} \phi_i \leq 1. \tag{6.78}$$

Then, for background noise density N_0, bandwidth W, and common user data rate R, the ratio of bit energy to interference density for the ith user[10] will be

$$\left(\frac{E_b}{I_0}\right)_i \geq \frac{\beta\phi_i S_{R_{1i}}/R}{\left(\sum_{j=1}^{J} S_{R_{ji}} + N_0 W\right)/W}. \tag{6.79}$$

Now suppose all users of the cell (or sector) are allocated the same E_b/I_0. As in the reverse link, we may want to vary this amount if a user is experiencing excessive multipath conditions. However, the variation will be minor and can be accounted for by assuming smaller $S_{R_{1i}}$ for that user. Then with $(E_b/I_0)_i = E_b/I_0$ for all i, we may invert (6.79) as an equality to obtain the relative allocation for the ith user,

$$\phi_i = \frac{E_b/I_0}{\beta W/R}\left(1 + \sum_{j=2}^{J} S_{R_{ji}}/S_{R_{1i}} + N_0 W/S_{R_{1i}}\right). \tag{6.80}$$

Now suppose we require a given value of E_b/I_0 (which, according to Section 4.4, should be on the order of 4 to 5 dB for the coherently demodulated forward link). Then this value will be sustainable for all users provided the fractional power allocations ϕ_i satisfy (6.78). Whenever this condition, and hence the desired E_b/I_0 level, cannot be met, the system will be deemed to be in an outage condition. We now proceed to evaluate and bound the outage probability. Recalling that the ith user is active only intermittently with probability given by (6.40) and that the number of

---

[10] This is a slight underbound because a small fraction, $\beta\phi_i$, of the power in $S_{R_{1i}}$ should not be included in the first term of the sum in the denominator.

users k_u is a Poisson random variable, we have

$$
\begin{aligned}
P_{out} &= \Pr\left(\sum_{i=1}^{k_u} \phi_i > 1\right) \\
&= \Pr\left[\sum_{i=1}^{k_u} \nu_i \left(1 + \sum_{j=2}^{J} S_{R_{ji}}/S_{R_{1i}} + N_0 W/S_{R_{1i}}\right) > \frac{\beta W/R}{E_b/I_0}\right],
\end{aligned}
\tag{6.81}
$$

where ν_i is the binary random variable defined by (6.40).

Generally the background noise (primarily of thermal origin) will be negligible compared to the total signal power (including all users' signals) received from all base stations. We can thus drop $N_0 W$ in (6.81). This is justifiable because base station transmitter power is not as limited as it is for mobile transmitters. Then, defining

$$
y_i = \sum_{j=2}^{J} S_{R_{ji}}/S_{R_{1i}}
\tag{6.82}
$$

and, as before,

$$
K_o = \frac{W/R}{E_b/I_0},
\tag{6.83}
$$

we have the Chernoff bound

$$
\begin{aligned}
P_{out} &= \Pr\left[\sum_{i=1}^{k_u} \nu_i(1 + y_i) > \beta K_o\right] \\
&< E \exp\left[\sum_{i=1}^{k_u} s\, \nu_i(1 + y_i) - s\beta K_o\right], \qquad s > 0.
\end{aligned}
\tag{6.84}
$$

Taking expectations with respect to k_u, ν_i, and y_i, where the latter are all mutually independent, we obtain

$$
\begin{aligned}
P_{out} &< \operatorname*{Min}_{s>0} E_{k_u} \prod_{i=1}^{k_u} E_{\nu_i} E_{y_i} e^{s\nu_i(1+y_i)} e^{-s\beta K_o} \\
&= \operatorname*{Min}_{s>0} E_{k_u}[\rho E_y e^{s(1+y)} + 1 - \rho]^{k_u} e^{-s\beta K_o} \\
&= \operatorname*{Min}_{s>0} \sum_{k_u=0}^{\infty} \frac{1}{k_u!} (\lambda/\mu)^{k_u} \{\rho[E_y e^{s(1+y)} - 1] + 1\}^{k_u} e^{-\lambda/\mu} e^{-s\beta K_o} \\
&= \operatorname*{Min}_{s>0} \exp\{\rho(\lambda/\mu)E_y[e^{s(1+y)} - 1] - s\beta K_o\}.
\end{aligned}
\tag{6.85}
$$

If we assume equal transmitted powers from all base stations, the power $S_{R_{ji}}$ received by the ith user from the jth base station is proportional to the inverse of the log-normally distributed attenuation of (6.1). Thus,

$$\frac{S_{R_{ji}}}{S_{R_{1i}}} = \frac{r_{1i}^m 10^{\zeta_{1i}/10}}{r_{ji}^m 10^{\zeta_{ji}/10}}.$$ (6.86)

r_{ji} is the distance from the jth base station to the ith user, and ζ_{ji} is the corresponding log-normal component of the attenuation. Further, by the definition (6.3), we may also express the sum of ratios as

$$y_i = \sum_{j=2}^{J} \frac{S_{R_{ji}}}{S_{R_{1i}}} = r_{1i}^m 10^{b\xi_{1i}/10} \sum_{j=2}^{J} r_{ji}^{-m} 10^{-b\xi_{ji}/10}.$$ (6.87)

Here, ξ_{ji} are independent with log-normal standard deviation $\sigma = 8$ dB. We also take $b = 1/\sqrt{2}$ and $m = 4$ and recall that

$$r_{1i}^m 10^{b\xi_{1i}/10} = \operatorname*{Min}_{j} r_{ji}^m 10^{b\xi_{ji}/10}.$$

The distribution of y_i, which depends both on relative distances and on the log-normally distributed variables, is not tractable analytically. However, it is straightforward to obtain the distribution by Monte Carlo simulation. We may assume that only base stations within the two concentric rings around the given cell transmit powers that are received at a discernible level by the given user. Thus, $J = 19$, including the given base station, the six base stations in the first ring, and the 12 base stations in the second ring. The variable y_i depends on the position of the ith user, which is assumed to be uniformly distributed in space. Thus, the distribution is averaged over all positions in the cell. However, by symmetry it is only necessary to perform simulations over the 30° right triangle shown in Figure 6.8.

In all, 100 trials were performed for each of 5,150 equally spaced points on the triangle. The result is the histogram of Figure 6.9. Using this histogram we may obtain numerically the expectation

$$E_y(e^{sy}) = \sum_{\{Y\}} e^{sY} \Pr(y = Y).$$ (6.88)

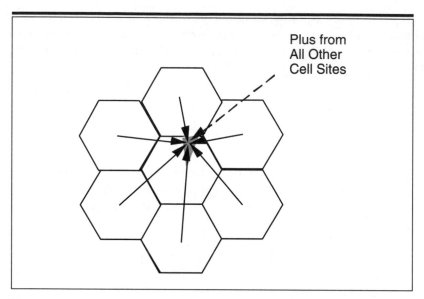

Figure 6.8   Forward link allocation geometry. ©1991 IEEE. "On the Capacity of a Cellular CDMA System" by K. S. Gilhousen, I. M. Jacobs, R. Padovani, A. J. Viterbi, L. A. Weaver, Jr., and C. E. Wheatley III, in *IEEE Transactions on Vehicular Technology*. Vol. 40, No. 2, pp. 303–312, May 1991.

From this the Chernoff bound of (6.85) can be computed, yielding the result[11] of the leftmost curve in Figure 6.10 for $\beta K_0 = 256$.

6.7.2 Soft Handoff Impact on Forward Link

As noted at the beginning of this section, there is considerable asymmetry between forward and reverse link conditions. Soft handoff, in particular, provides a considerable gain in *reverse link* performance with virtually no interference drawbacks, because the second cell simultaneously receiving useful signals from the mobile user would be receiving that energy in any case, without advantage if it were not in soft handoff. Establishing soft handoff for the *forward link* is another matter. The second cell base station must now transmit to the given user the same signal as the first cell. Hence, it must now *transmit* energy that is therefore no longer available for allocations to other users of that second cell.

---

[11] Although typically $\beta = .8$ (or -1 dB), we take E_b/I_0 to be 1 dB less than for the reverse link, so that βK_0 for the forward link is approximately the same as K_0 for the reverse link. The factor $1 + g$, in the abscissa of Figure 6.10, will be explained in the next subsection. It is understood that for a sectored cell, all attributes apply to individual sectors of the base station.

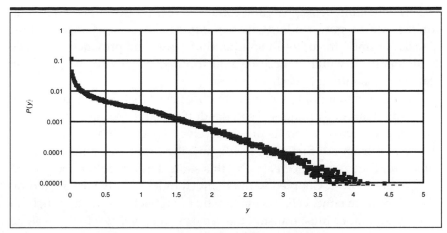

Figure 6.9 Histogram of other-cell forward relative interference (mean = 0.51). ©1991 IEEE. "On the Capacity of a Cellular CDMA System" by K. S. Gilhousen, I. M. Jacobs, R. Padovani, A. J. Viterbi, L. A. Weaver, Jr., and C. E. Wheatley III, in *IEEE Transactions on Vehicular Technology*, Vol. 40, No. 2, pp. 303–312, May 1991.

Consider now a user in soft handoff. Under the best of circumstances, where propagation losses from each base station to the user are the same, the power levels received from the two base stations are nearly equal. Then each cell would need to transmit less power than before, since the mobile receiver coherently combines what is received from both transmitters. This could be ideally 3 dB less for each transmitter if it received

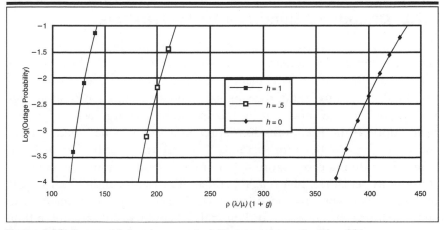

Figure 6.10 Forward link outage probability upper bounds; $\beta K_0 = 256$.

exactly equal unfaded signals and combined them coherently, so that total interference or power allocation requirements were unchanged. Thus, signals destined for users in soft handoff would be provided the same total energy. It would simply be redistributed equally to the two base stations transmitting to each such user.

This ideal situation, however, is almost never met. In fact, one of the major qualitative advantages of soft handoff is that the second base station can be made active when its power received by the mobile user is well below that from the first (e.g., 6 dB), and the first base station is maintained until its power decreases by this same amount below the second base station power. Coupled with the much more rapidly varying fading conditions from either cell, it is evident that for effective soft handoff, both cells' base stations must transmit the same power level as when one alone is transmitting to the user. As a result, users in soft handoff contribute approximately twice as much interference as users that are not in soft handoff.

Assume then that, in a uniformly loaded multiple cell system, a fraction $g < 1$ of all users is in soft handoff. Also assume that both base stations involved for each such user allocate essentially the same fraction ϕ_i to that user. Effectively, then, the arrival rate of users per base station involved in the outage probability derivation of (6.81) to (6.88) rises from λ to $\lambda(1 + g)$. Thus, in the Chernoff bound of (6.85), λ is increased by the factor $1 + g$, and in the resulting graph of Figure 6.10, the abscissa becomes $\rho(\lambda/\mu)(1 + g)$.

The conclusion is obvious: The soft handoff condition thresholds, where new base stations enter the condition and old ones leave it, must be selected so that g is kept relatively low (on the order of $\frac{1}{4}$ to $\frac{1}{3}$).

6.7.3 Orthogonal Signals for Same-Cell Users

Throughout this text we have assumed that independent pseudorandom sequences are used to spread each user's signal, whether in the reverse or the forward link. We now consider an alternative approach for the forward link only: a hybrid of the slot assignment concept and random signaling. Rather than randomizing the spread signals for each individual user, suppose the pseudorandom spreading signal is assigned only to individual base stations (or sectors). Thus, referring to the forward link transmitter diagram of Figure 4.2, a single pseudorandom sequence is introduced per carrier phase, $a_n^{(I)}(0)$ and $a_n^{(Q)}(0)$, after the individual signals are combined. Now instead of individual pseudorandom sequences to separate the users, we assign orthogonal slots but the slots are not in time or frequency, as in conventional systems. Rather, a unique orthogonal sequence is provided to each. This is most easily done by assigning one of $k_u \leq N$ Hadamard-Walsh functions of duration N chips (or a multiple thereof) to spreading each symbol of each user, as shown in Figure 6.11. Orthogonalization of same-cell users requires a corresponding minor modification of the mobile receiver as well. When the base station receives and recognizes a newly accessing user, it must assign it a Hadamard–Walsh function. Then, while the base station transmitter multiplies the coded data sequence by this function to spread and orthogonalize, the receiver demodulator must similarly multiply by the same sequence, as well as by the cell's (or sector's) pseudorandom sequences, to remove the spreading before BPSK demodulation. In soft handoff, the new cell also must assign a Hadamard–Walsh function for transmission to the mobile user, which generally will be different from the one used by the first cell (since that one may already have been assigned). This means that each

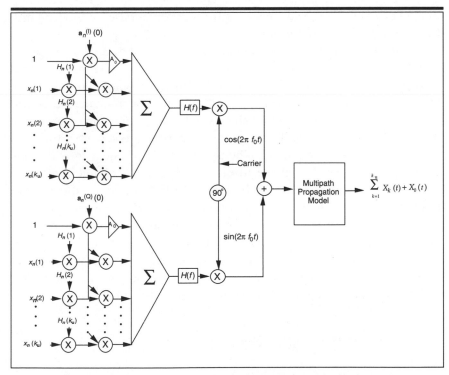

Figure 6.11 Multiuser (base station) modulator employing orthogonal Hadamard–Walsh multiplexing. $H_n(k)$ = Hadamard–Walsh sequence for kth user; $n = 1, 2, \ldots , N; k \leq k_u \leq N$.

demodulator finger associated with the second cell will use the new function for removing the spreading from that base station.

For a spreading factor W/R, when the users' data sequences are coded with an FEC code of rate $r < 1$, then the number of orthogonal Hadamard–Walsh "slots" available and hence the number of users supportable without soft handoff is

$$k_u \le (W/R)r. \tag{6.89}$$

With soft handoff, this is reduced by the factor $1 + g$, where g is the fraction of users in soft handoff. Note that this is a hard limit, as is the case for all slotted systems. We may, however, double this value by modulating k_u users on the in-phase (I) carrier and an additional k_u users on the quadrature (Q) carrier. Orthogonality is obviously maintained at the mobile receiver by coherent demodulation using the received pilot. All combined signals on both I and Q carriers are randomized by inserting the base-station specific pseudorandom sequences prior to PSK modulation of either carrier. Thus, transmissions from different base stations are mutually independent random processes.

Ideally, if the same-cell users are orthogonalized, they no longer appear as interference to one another. Thus, in (6.79), the denominator would start with $j = 2$. Similarly, in (6.80) and all the following, the unity term would disappear. Unfortunately, conditions are not so favorable: Multipath will introduce interference that is not orthogonal among same-cell users. In fact, if the multipath components are separated by more than one chip time, the base-station–specific pseudorandom sequence will make them appear as random noise to one another. For example, with two equal-strength paths separated by one chip time or more, the relative interference will be half (3 dB below) what it would be without orthogonalization. Both less favorable and more favorable situations will arise. In any case, suppose the same-cell interference is reduced by orthogonalization to a factor $h < 1$. Then in all expressions from (6.80) onward the unity term is replaced by h. Then, accounting both for orthogonalization and for soft handoff of a fraction g of users, (6.85) is modified to become

$$P_{out} < \underset{s>0}{\text{Min}} \exp\{(1 + g)\rho(\lambda/\mu)\, E_y[e^{s(h+y)} - 1] - s\beta K_0\}. \tag{6.90}$$

The result is shown in Figure 6.10, for $h = 0, .5$, and 1, plotted as a function of $(1 + g)\rho(\lambda/\mu)$. We can see by comparison with Figure 6.4 that for

$h \geq 0.5$ and $g < f$, the forward link capacity limit normally exceeds that of the reverse link.

Finally, given the moderate improvement achievable through orthogonalization, we might consider employing it on the reverse link as well. This poses a difficulty with timing synchronization. Orthogonalization is possible on the forward link because all users emanating from the same transmitter have exactly synchronized common timing. On the reverse link, the multiple users are completely uncoordinated. To synchronize chip timing would require a closed-loop timing control process, similar to that used for power control. The tracking accuracy would need to be on the order of a small fraction of a chip time. Such a complex procedure may not be warranted for only a moderate improvement in performance. Furthermore, if the reverse link employs M-ary noncoherent modulation, which already expands bandwidth, the additional bandwidth expansion for orthogonalization may exceed the total available expansion factor W/R. Of course, the reverse link might employ binary PSK or differential PSK modulation. But then, to maintain performance, coherent demodulation of the reverse link would also be desirable, which probably requires an unmodulated pilot signal or unmodulated time-segment for each user.

6.8 Interference Reduction with Multisectored and Distributed Antennas

A basic assumption made throughout this chapter, as stated in Section 6.4, is that users are uniformly distributed throughout each of many cells over an arbitrarily large two-dimensional region. In addition, we have assumed idealized sectored antennas with a square cutoff pattern, whose receive and transmit gain is unity over its beamwidth and zero elsewhere. Thus, all sectors are completely disjoint, and the sectored antennas operate independently of the other sectored antennas of the same cell. For example, with three-sectored antennas, each covering exactly $2\pi/3$ radians, the Erlang capacity is the same for each sector. Hence, the cell's capacity is three times that of a cell with an omnidirectional antenna at the base station. This follows from the fact that with a uniform distribution of users, the other-cell interference is simply the total interference from users controlled by other sectored cell base stations that also lie in the $2\pi/3$ radian sector viewed from the given base station's sectored antenna. Ideally this improvement could be made N-fold by employing N such hypothetical $2\pi/N$ radian beamwidth antennas.

Maintaining the assumption of a uniform density of users, we next consider a more realistic base station with sectored antennas whose patterns overlap. These antennas may be separately implemented, or they may be the multiple outputs of a phased array. In the latter case, a set of N antennas (possibly omnidirectional) are combined linearly to generate a set of $M \leq N$ outputs, each corresponding to a beam whose width is approximately $2\pi/M$ radians.[12] In this case, we may slightly modify the interference calculation of Section 6.4 by including the effect of the antenna gains. This has the effect of dividing the other-user omnidirectional interference term by G_A, the antenna gain, rather than by the number of sectors, N. G_A is defined as the angular pattern amplitude in the direction of the desired user divided by the average pattern amplitude over 2π radians. The resulting cell capacity (of both forward and reverse links) is obtained as before for an omnidirectional cell, but with W/R multiplied by G_A. Alternatively, the sector capacity is obtained by multiplying the previous (ideal) sector capacity by G_A/N.

There still remain a number of issues regarding multipath combining. If the antenna sector beamwidths are relatively narrow (high gain), multiple paths may be received at the base station through different sectors. Combining these paths requires summing different element outputs (coherently or noncoherently, depending on circumstances described in Chapter 4). With lack of reciprocity for forward and reverse links, however, multiple element combining may not produce the same transmission gain as the reception gain it produces.

Another important consideration is that of homogenous user density. In most areas, the assumption is not valid. In extreme cases, such as along a seacoast bordered by a mountain range, the user population may be almost linearly distributed along a narrow strip. Obviously, in such cases, sectored antennas are virtually useless because all other-user interference resides in a relatively narrow beamwidth. Then, a way to increase capacity is to place microcells (with smaller diameters and lower antenna heights) along the strip, but control them centrally at one base station [Lee, 1994]. The base station performs all processing, including power control; this, of course, requires a wideband interconnection of the antennas. Such an approach is even more useful indoors, where propagation may be limited to corridors and individual rooms. Here the use of distributed antennas with many small elements may be the best way to provide coverage. Additionally, if each antenna element's transmission is delayed relative to its neighbors by an interval greater than $1/W$, the signal received at the mobile from two or more elements transmitting the same signal can be combined to advantage by a rake receiver. This is another

---

[12] Because of nonuniform propagation of wavefronts, this may require adaptive linear combination of the antenna elements [Naguib *et al.*, 1994].

example of soft handoff, described and analyzed in Sections 6.4 and 6.5. All of the foregoing alternatives are possible because of the universal frequency reuse feature of spread spectrum multiple access.

The design of antenna systems, whether by sectoring, with arrays, or with distributed elements, provides perhaps the greatest opportunity for increasing capacity for any future advances in cellular and personal communication networks. Tailoring the antenna technique to the physical environment and population distribution, injecting and extracting signal power to and from where it is needed, is an important means of increasing capacity through spatial interference management.

6.9 Interference Cancellation

For a single-cell multiple access system, assuming unfaded transmission for each user over a common additive white Gaussian noise channel, information theory can be used to demonstrate that interference from same-cell users can be completely eliminated by a process of successive cancellation of interfering users [Wyner, 1974; Viterbi, 1990]. The resulting capacity of the multiple user channel is given by the classical AWGN formula [Shannon, 1949], with rate referring to the sum of the rates of all users and signal power being the sum of the received powers (at the base station) from all users. Though the proof of this result is constructive, meaning that the method for achieving the result is specified, there are several aspects of the technique that will always render it impractical:

(a) an arbitrarily long and powerful FEC code is required;

(b) arbitrarily long processing delays are involved;

(c) reception is unfaded, or fading is very slow compared to a frame duration;

(d) all received users must be processed together to recover any one user.

All four of these conditions can be relaxed, with some compromise in performance, to achieve a reasonable level of complexity and delay. However, the resulting performance will most likely be worse than can be achieved by applying other simple, more robust techniques, most notably the antenna design methods discussed in the last section.

We will not consider specific interference cancellation techniques for the reverse link, as has been done by several authors [Viterbi, 1990; Yoon *et al.*, 1992; Dent, 1993]. Instead, we establish only some upper bounds on the maximum improvement of interference cancellation in a cellular system. Our premise is that each base station processes all the users it controls, but none of those controlled by other base stations. For simplicity and to obtain explicit results, we also assume the idealized condition of unfaded transmission (no multipath, or at least well-separated paths of known amplitudes) and coherent reception. Then if perfect interference cancellation were possible (which would require the aid of a "genie"), the interference density would be reduced from $(1 + f)I_0$ to fI_0, where f is the relative other-cell interference as computed in Section 6.4. Thus, not surprisingly, this upper bound on interference cancellation gain for the reverse link cellular system, $(1 + f)/f$, is the same as the maximum benefit of using orthogonal (Hadamard) sequences for user separation on the forward link—or, for that matter, on the reverse link, if the additional features and procedures are implemented on the latter to provide orthogonality and coherence as described at the end of Section 6.7.3. We show this genie-aided interference cancellation limit on improvement as the top curve in Figure 6.12. Thus, for example, if $f = 0.6$, the improvement factor is 2.67, or 4.3 dB.

We now show a less optimistic assessment of the interference cancellation improvement for a system that employs a powerful FEC code. The classical Shannon formula provides an upper bound on capacity for a cellular system with interference cancellation. The total same-cell user received power from all users of the cell or sector is $k_u P_S$. P_S is the equal (power-controlled) received power per user of the given cell. Also, as

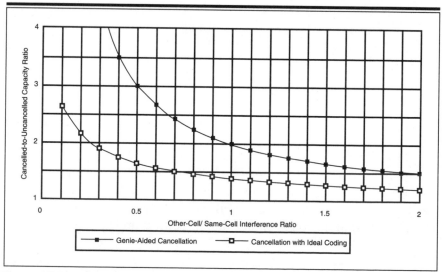

Figure 6.12 Maximum capacity improvement by same-cell interference cancellation.

assumed, the total (uncancellable) other-cell (or sector) interference is fk_uP_s. Then the maximum total rate received by a base station from its users each transmitting at rate R is bounded by

$$k_uR < W\log_2\left(1 + \frac{k_uP_s}{fk_uP_s}\right),\qquad(6.91)$$

or

$$k_u < \frac{W}{R}\log_2\frac{\ln(1+1/f)}{\ln 2}.\qquad(6.92)$$

If this were not so, and this limit could be exceeded, then treating fk_uP_s as background Gaussian noise, N_0W, a single user of the cell could break up its signal into k_u additive components, with total power k_uP_s. Then, using interference cancellation, it could operate above the total rate given by the right side of (6.91), thus exceeding Shannon capacity.

To provide a reasonable comparison to the case without cancellation, we assume in the latter that each user is operating independently, with the same idealized FEC coding operating at the Shannon capacity for its (uncancelled) channel. Then in this case, with $W \gg R$, each user's E_b/I_0 approaches $\ln 2$. Using the standard formula (1.5), without gain factors that were also not assumed with cancellation,[13]

$$k_u < \frac{W/R}{(E_b/I_0)(1+f)}.$$

With $E_b/I_0 = \ln 2$, we obtain

$$k_u < \frac{W/R}{\ln 2(1+f)}.\qquad(6.93)$$

Then, taking the ratio of (6.92) and (6.93) for these idealized cancelled and uncancelled cases, we obtain

$$\frac{k_u(\text{interference cancellation})}{k_u(\text{uncancelled})} = (1+f)\ln(1+1/f),\qquad(6.94)$$

---

[13] In keeping with the assumptions leading to (6.92), we assume no additional features other than coherent reception with ideal coding, and we take no reduction for variability of power control or traffic intensity.

which is shown as the lower curve in Fig. 6.12. We note that for $f = 0.6$, the ratio is only 1.57 (or 2.0 dB), as compared with 2.67 (or 4.3 dB) for the genie-aided case. We may conclude that when powerful FEC is employed, the additional advantage of interference cancellation is considerably reduced from the "genie-aided" limit. Besides the obvious fact that interference cancellation will involve some processing delay, especially if decoding is required before cancellation, a more subtle issue is its interaction with the closed-loop power control process.[14]

Finally, we note that since interference cancellation generally requires coherent reception, making the same-cell user signals orthogonal on the reverse link with time synchronization, as discussed in Section 6.7.3, may be easier and will provide better performance than interference cancellation.

6.10 Epilogue

This text is an unorthodox treatment of digital communication theory, stressing the physical layer of multiple user, multiple cell, communication networks. Among the unconventional topics that we have introduced, the foremost are

(a) *universal* frequency and time occupancy, as contrasted with the frequency or time *segmentation* of conventional frequency-and time-division multiple access;

(b) the role of power control in improving the efficiency of two-way systems;

(c) soft handoff among multiple nodes in a cellular network, particularly as an extension of constructive combining of multipath propagation, by means of a rake receiver;

(d) the central role of low-rate forward error correction in enhancing terrestrial transmission, which is more bandwidth-limited than power-limited;

(e) increased network capacity through distributed, not centrally controlled, adaptive processing to exploit source rate and spatial interference variabilities.

---

[14] Among other issues, operating at capacity approaching the Shannon limit (6.91), with equal rate users, requires that the received user powers be unequal [Viterbi, 1990], which will complicate the power control process.

We repeat our comment in the preface that though it is strongly motivated by applications, this text has concentrated on principles rather than detailed methods. It has been our goal to provide an understanding of the "why" of the implementation techniques of a CDMA system rather than the details of "how." As a side effect, we have avoided any but the most general comparisons with slotted (FDMA and TDMA) systems. This is partly because comparisons are invidious, and partly because any comparison must be either with existing and, by definition, obsolescent systems, or projected, and hence unproven systems.

Aside from brief remarks in Chapters 4, 5, and 6, and some common elementary analytic techniques, this book has not dealt with information theory. Nonetheless, we should acknowledge the fundamental relationships between spread spectrum communications and Shannon's information theory. A superficial discussion of the conceptual antecedents and theoretical justification provided by information theory to spread spectrum digital communication appeared in recent articles by the author [Viterbi, 1991, 1993b, 1994]. Deeper understanding of this relationship is found in multiuser information-theoretic papers and texts [Wyner, 1974; Cover and Thomas, 1991]. Though the latter do not illuminate the path toward practical system realization, we firmly believe that, as in the past, the next significant advance will depend heavily on such foundations. It is our hope that this text will serve as a step along the path of full exploitation of theoretical concepts by means of ever-improving implementation technologies.

REFERENCES AND BIBLIOGRAPHY

Bertsekas, D., and Gallager, R. (1987). *Data Networks*. Prentice-Hall, Inc., Englewood Cliffs, New Jersey.

Brady, P. T. (1968). "A Statistical Analysis of On–Off Patterns in 16 Conversations," *Bell Syst. Tech. J.* **47**, 73–91.

Cover, T. M., and Thomas, J. A. (1991). *Elements of Information Theory*. J. Wiley, New York.

Dent, P. W. (1993). "CDMA Subtractive Demodulation," U.S. Patent #5,218,619, issued June 8.

Feller, W. (1957). *An Introduction to Probability Theory and Its Applications*, Vol. I, 2nd Ed. Wiley, New York.

Forney, G. D., Jr. (1970a). "Coding and Its Application in Space Communications," *IEEE Spectrum* **7**, 47–58.

Forney, G. D., Jr. (1970b). "Convolutional Codes I: Algebraic Structure," *IEEE Trans. Inform. Theory* **IT-16,** 720–738.

Gallager, R. G. (1968). *Information Theory and Reliable Communication*. Wiley, New York.

Gardner, F. M. (1968). *Phaselock Techniques*. Wiley, New York.

Gilhousen, K. S., Jacobs, I. M., Padovani, R., Viterbi, A. J., Weaver, L. A., and Wheatley, C. E. (1991). "On the Capacity of a Cellular CDMA System," *IEEE Trans. on Vehicular Technology* **40**(2), 303–312.

Golomb, S. W. (1967). *Shift Register Sequences*. Holden-Day, San Francisco. Revised edition (1982), Aegean Park Press, Laguna Hills, California.

Heller, J. A. (1968). "Short Constraint Length Convolutional Codes," Jet Propulsion Laboratory, Space Programs Summary 37–54, Vol. III, 171–177.

Helstrom, C. W. (1968, 1960). *Statistical Theory of Signal Detection*. Pergamon, London.

Jakes, W. C., Jr. (Ed.) (1974). *Microwave Mobile Communications.* J. Wiley & Sons, New York.

Lee, W. C. Y. (1989). *Mobile Cellular Telecommunications Systems.* McGraw-Hill, New York.

Lee, W. C. Y. (1994). "Applying the Intelligent Cell Concept to PCS," *IEEE Trans. on Vehicular Technology* **43**(3), 672–679.

Lindsey, W. C., and Simon, M. K. (1973). *Telecommunication Systems Engineering.* Prentice-Hall, Englewood Cliffs, New Jersey.

Marcum, J. I. (1950). "A Table of Q-Functions," *Rand Corp. Report,* RM-339, January.

Mason, S. J. (1956). "Feedback Theory—Further Properties of Signal Flow Graphs," *Proc. IRE* **41**, 920–926.

Massey, J. L., and Sain, M. (1968). "Inverses of Linear Sequential Circuits," *IEEE Trans. Comput.* **C-17**, 330–337.

McEliece, R., Dolinar, S., Pollara, F., and Van Tilborg, H. (1989). "Some Easily Analyzable Codes," presented at Proceedings of the Third Workshop on ECC, IBM Almaden Research Center, September.

Naguib, A. F., Paulraj, A., and Kailath, T. (1994). "Capacity Improvement with Base-Station Antenna Arrays in Cellular CDMA," *IEEE Trans. on Vehicular Technology* **43**(3), 691–698.

Neyman, J., and Pearson, E. S. (1933). "On the Problem of the Most Efficient Tests of Statistical Hypotheses," *Phil. Trans. Roy. Soc. London, Series A* **231**, 289–337.

Padovani, R. (1994). "Reverse Link Performance of IS-95 Based Cellular Systems," *IEEE Personal Communications Magazine* **1**, Third Quarter, 28–34.

Peterson, W. W., and Weldon, E. J., Jr. (1972). *Error-Correcting Codes,* 2nd Ed. MIT Press, Cambridge, Massachusetts.

Plotkin, M. (1960). "Binary Codes with Specified Minimum Distance," *IRE Trans. Inform. Theory* **IT-6**, 445–450.

Polydoros, A., and Simon, M. K. (1984). "Generalized Serial Search Code Acquisition: The Equivalent Circular State Diagram Approach," *IEEE Trans. Commun.* **COM-32**, 1260–1268.

Price, R., and Green, P. E., Jr. (1958). "A Communication Technique for Multipath Channels," *Proc. IRE* **46**, 555–570.

Proakis, J. (1989). *Digital Communications,* 2nd Ed. McGraw-Hill, New York.

Pursley, M. B. (1977). "Performance Evaluation for Phase-Coded Spread-Spectrum Multiple Access Communication—Part I: System Analysis," *IEEE Trans. Comm.* **COM-25,** 795–799.

Schwartz, M., Bennett, W. R., and Stein, S. (1966). *Communication Systems and Techniques.* McGraw-Hill, New York.

Shannon, C. E. (1949). "Communication in the Presence of Noise," *Proc. IRE* **37,** 10–21.

Simon, M. K., Omura, J. K., Scholtz, R. A., and Levitt, B. K. (1985) *Spread Spectrum Communications, Vol. I, II, III.* Computer Science Press, Rockville, Maryland.

Spilker, J. J., Jr. (1963). "Delay-lock Tracking of Binary Signals," *IEEE Trans. Space Electr. and Telem.* **SET-9**(1), 1–8.

Steele, R. (Ed.) (1992). *Mobile Radio Communications.* Pentech Press, London.

Turin, G. L. (1980). "Introduction to Spread Spectrum Antimultipath Techniques and their Application to Urban Digital Radio," *Proc. IEEE,* **68,** 328–354.

Vijayan, R., Padovani, R., Wheatley, C., and Zehavi, E. (1994) "The Effects of Lognormal Shadowing and Traffic Load on CDMA Cell Coverage," Qualcomm Incorporated Internal Memorandum.

Viterbi, A. J. (1966). *Principles of Coherent Communication,.* McGraw-Hill, New York.

Viterbi, A. J. (1967a). "Error Bounds for Convolutional Codes and an Asymptotically Optimum Decoding Algorithm," *IEEE Trans. Inform. Theory* **IT-13,** 260–269.

Viterbi, A. J. (1967b). "Orthogonal Tree Codes for Communication in the Presence of White Gaussian Noise," *IEEE Trans. Commun. Tech.* **COM-15,** 238–242.

Viterbi, A. J. (1971). "Convolutional Codes and Their Performance in Communication Systems," *IEEE Trans. Commun. Tech.* **COM-19,** 751–772.

Viterbi, A. J. (1990). "Very Low Rate Convolutional Codes for Maximum Theoretical Performance of Spread-Spectrum Multiple-Access Channels," *IEEE J. on Selected Areas in Communication* **8,** 641–649.

Viterbi, A. J. (1991). "Wireless Digital Communication: A View Based on Three Lessons Learned," *IEEE Communications Magazine,* 33–36.

Viterbi, A. J. (1993a). "Method and Apparatus for Generating Super-Orthogonal Convolutional Codes and the Decoding Thereof," U.S. Patent #5,193,094, issued March 9.

Viterbi, A. J. (1993b). "Overview of Mobile and Personal Communication," in *Modern Radio Science* (H. Matsumoto, Ed.). Oxford University Press, Oxford, U.K.

Viterbi, A. J. (1994). "The Orthogonal-Random Waveform Dichotomy for Digital Mobile Personal Communication," *IEEE Personal Communications Magazine*, 2–8, First Quarter.

Viterbi, A. J., and Omura, J. (1979). *Principles of Digital Communication and Coding.* McGraw-Hill, New York.

Viterbi, A. J., and Padovani, R. (1992). "Implications of Mobile Cellular CDMA," *IEEE Communications Magazine*, 38–41, December.

Viterbi, A. J., Viterbi, A. M., and Zehavi, E. (1993). "Performance of Power-Controlled Wideband Terrestrial Digital Communication," *IEEE Transactions on Communications* **41**(4), 559–569.

Viterbi, A. J., Viterbi, A. M., Gilhousen, K. S., and Zehavi, E. (1994). "Soft Handoff Extends CDMA Cell Coverage and Increases Reverse Link Capacity," *IEEE J. Selected Areas in Communications* **12**(8), 1281–1288.

Viterbi, A. M., and Viterbi, A. J. (1993). "Erlang Capacity of a Power Controlled CDMA System," *IEEE Journal on Selected Areas in Communications* **11**(6), 892–900.

Wolf, J. K., Michelson, A., and Levesque, A. (1982). "On the Probability of Undetected Error for Linear Block Codes," *IEEE Trans. Comm.* **COM-30**, 317–324.

Wozencraft, J. M., and Jacobs, I. M. (1965). *Principles of Communication Engineering.* Wiley, New York.

Wyner, A. D. (1974). "Recent Results in the Shannon Theory," *IEEE Trans. Inform. Theory* **IT-20**, 2–10.

Yoon, Y. C., Kohno, R., and Imai, H. (1992). "A Spread-Spectrum Multi-access System with a Cascade of Co-channel Interference Cancellers for Multipath Fading Channels," *IEEE Second International Symposium on Spread Spectrum Techniques and Applications (ISSSTA'92)*, 87–90, Yokohama, Japan.

Zehavi, E., and Viterbi, A. J. (1990). "On New Classes of Orthogonal Convolutional Codes," Bilkent International Conference in New Trends in Communication, Control, and Signal Processing, Ankara, Turkey.

INDEX